# INFECTIOUS FUNGI

RECENT ADVANCES IN PLANT PATHOLOGY SERIES

# INFECTIOUS FUNGI

K. M. CHANDNIWALA

ANMOL PUBLICATIONS PVT. LTD.
NEW DELHI-110 002 (INDIA)

**ANMOL PUBLICATIONS PVT. LTD.**
**Regd. Office:** 4360/4, Ansari Road, Daryaganj,
New Delhi-110002 (India)
Tel.: 23278000, 23261597, 23286875, 23255577
Fax: 91-11-23280289
*Email:* anmolpub@gmail.com
*Visit us at:* www.anmolpublications.com

**Branch Office:** No. 1015, Ist Main Road, BSK IIIrd Stage
IIIrd Phase, IIIrd Block, Bangalore-560 085 (India)
Tel.: 080-41723429 • Fax: 080-26723604
*Email:* anmolpublicationsbangalore@gmail.com

*Infections Fungi*



First Edition, 1996
Reprint, 2010
ISBN 81-7488-276-6



PRINTED IN INDIA

---

Printed at Mehra Offset Press, Delhi.

# Preface

The present title has been intended primarily for the use of students enrolled in advanced undergraduate and postgraduate courses in plant pathology. The need of a textbook in this field was a pressing one since not only did the students find literature on the diseases and other pathological conditions difficult to obtain and to access, but no reasonably complete treatment of the subject from the student's stand point was to be had. To fill this need is the first objective of this title.

Considering the field as a whole, however, it was realised that in addition to a textbook for classroom use, the subject was also greatly in need of a reference work. The pathologists and the research worker in biology have not had an available guide to all phases of the subject of plant pathology as such for most of the information the research worker has been faced to consult are widely scattered in often almost inaccessible articles or incomplete accounts in larger volumes. Accordingly, the authors, turned their attention to the preparation of the present book, which deals specially with plant pathology, i.e. with the microbial diseases as well as with certain amicrobial deseases, injuries and abnormalities. The task has not been easy one.

Attempts have been made to present uptodate but brief introduction to principles underlying various stages in the development of plant diseases with special emphasis on their control. The language of the book has been painfully kept as simple and lucid style as possible. The subject matter has been interwoven for keeping up interest and curiosity of readers.

There can be no claim to originality except in the manner of treatment and much of the information has been obtained from the books and scientific journals available in different libraries.

Constructive criticism and suggestion for improvement of the book will be thankfully acknowledged.

**Author**

# Contents

# 1

# INFECTIOUS FUNGI

## FUNGOUS INFECTIONS ON INSECTS
## (Mycoses)

As far as the disease of insects are concerned, it seems only natural that the first type of infection definitely identified as to its microbial nature was that caused by fungi. This relationship was suspected about the beginning of the nineteenth century. The conspicuous presence of a mycelium and fruiting bodies on the cadavers of dead silkworms enabled Bassi de Lodi, in 1835, to recognize the similarity between the infecting fungus and the well-known bread molds and other saprophytic fungi. As further observations were made it was discovered that some of the fungi infecting insects were obligate parasites, others were semiparasites, and still others were common saprophytic species which, under certain conditions, were able to cause a frank infection in susceptible insects. A confusing factor was the frequent development of a saprophytic fungus in the body of an insect killed by an agent other than the fungus. Such secondary invaders were often mistaken for the primary cause of the condition, and as a result the invasive properties of some of the fungi described in early reports are in doubt and these organisms probably are not pathogenic agents in the true sense of the term.

There are certain fungus-insect relationships that could not properly be considered in the category of disease (e.g., the *Septobasidium* fungi and the Laboulbeniaceae), and others that are of a definitely mutualistic association (e.g., the ambrosia fungi of termites and wood-boring beetles). These relationships are therefore outside the limits of the usual concept as to what constitutes a fungous disease.

*Nature of the Fungi Concerned.* Fungi constitute a part of that subdivision of the plant kingdom known as Thallophytes, those organisms which grow in irregular plant masses (thalli) not differentiated into roots, stems, and leaves as in higher plants. The term "fungi" is a general one and may be meant to include several types of organisms that are sometimes conveniently differentiated by the terms "molds," "yeasts," "actinomycetes," "bacteria," etc. Commonly, however, the term "fungi" refers to that group of nonchlorophyll-containing organisms which usually possess the filamentous vegetative structure known as "mycelium." Or, stating it another way, the word "fungi" usually refers specifically to the molds and yeasts, which comprise the Eumycetes.

Representatives of the fungi known to cause infection and disease a insects ("*fungi Entomogeni*") are included in each of the four classes: Phycomycetes, Ascomycetes, Basidiomycetes, and Deuteromycetes (Fungi Imperfecti). Of these, the class basidiomycetes contains the fewest species characteristically pathogenic for insects. Because, under certain conditions, some groups of yeasts produce asci and ascospores, these micro-organisms are usually included with the Ascomycetes.

For the sake of convenience we shall treat the mycoses of insects more or less according to the class to which the causative fungus belongs. There is some overlapping between the fungi belonging to the class Ascomycetes and those placed in the class Deuteromycetes. The latter group are sometimes called "Fungi Imperfecti" because they are fungi with incomplete or incompletely known life cycles. In most cases they are believed to be imperfect (asexual) stages of Ascomycetes.

In passing, mention should be made of the fact that certain artifirms, foreign bodies, or pseudofungi occasionally occur on insects and are sometimes confusing to one examining the specimens for fungi. Certain seeds and pollinia, for example, frequently are found attached to the integuments of insects in a manner closely resembling certain fungi such as Laboulbeniaceae and *Cordyceps*. Usually these foreign nonparasitic bodies may be detached mechanically from the insect rather easily, although sometimes they cling as tenaciously as might be expected of a fungus. Pollinia may be attached by adhesive discs, which will ordinarily become detached from the insect if the specimen is moistened with a drop of alcohol. Some orchid pollinia, with their stalks of varying lengths and pollen masses at the apex, are similar in appearance to some of the stalked fungi but become detached readily upon the application of alcohol. The pollen of some plants covers insects (such as wasps) visiting their flowers with a white powder that is sometimes mistaken for a fungus, but this can usually be detected upon microscopic examination.

## PHYCOMYCETE INFECTIONS

Phycomycetes ("algal fungi") constitute a large and diverse group living in various habitats; some are aquatic, some amphibious, some terrestrial. They are characterized by coenocytic (nonseptate and multinucleate) mycelium and endogenous asexual spores. (Septations commonly are formed where the reproductive structures are delimited; they may be formed in old hyphae of certain genera and in young thalli of certain other genera. From the standpoint of insect infections, at least four orders of the class Phycomycetes contain entomogenous members: Entomophthorales, Mucorales, Blastocladiales, and Chytridiales. As the name would imply, the most important of these is the order Entomophthorales, which is generally considered to consist of a single family, Entomophthoraceae (Empusaceae of some authors), which in turn may be divided into five and possibly six genera. Of these; the genera *Empusa, Entomophthora*, and *Massospora* are composed primarily of entomogenous species.

### Infections Caused by Species of *Empusa* and *Entomophthora*

The literature records a relatively large number of infections of insects caused by species of *Empusa* and *Entomophthora*. Approximately 50 species of these fungi are known to occur in the United States. Some of these have been studied in considerable detail whereas others have received only taxonomic descriptions. In the main, their life histories are similar, and considerable general information may be gained by a close consideration of a few of the better known species.

The proper use of the generic names *Empusa* and *Entomophthora* is frequently a difficult procedure, and unless the student has an understanding of the historical aspect of these names he is likely to become somewhat confused. No final official ruling has been made on this particular nomenclatorial problem, and for the present one can only follow the usage generally employed by the leading authorities in the field. The validity of the name *Empusa*, erected by Cohn in 1855 for a fungous parasite (*Empusa muscae* Cohn) of the housefly, has been challenged by some writers because of its having been preoccupied for a genus of orchids. It was for this reason that Fresenius (1856) proposed the name *Entomophthora* to take the place of *Empusa* for the fungus. This was followed by the indiscriminate use of both names until Brefeld (1877) and Nowakowski (1884) separated them as two distinct genera, thereby recognizing the validity of the name *Empusa*. Other mycologists subsequently employed it without question. In his treatise on the Entomophthoreae of the United States, Thaxter (1888) concluded that the name had sufficient weight of authority to make it acceptable. He further pointed out that the orchidaceous genus *Empusa* is now placed as a synonym and hence seems unlikely to cause confusion.

The generic name *Entomophthora* was considered by Thaxter to be of subgeneric value in the genus *Empusa*. Since then, for reasons to be pointed out later, there has been a tendency to revert to the proposals of Brefeld and Nowakowski and to give *Entomophthora* full generic rank along with *Empusa*. The genus *Lamia*, erected by Nowakowski and considered by him to be intermediate between the two, has not found general acceptance. Unfortunately, a clear and final distinction between some of the species of *Empusa* and those of *Entomophthora* has not been made. It appears probable, however, that the two names will be used separately and that with further detailed studies two valid genera will be recognized.

Nature of *Empusa* and *Entomophthora* Infections. Among the various species of *Empusa* and *Entomophthora* there is some variation as to the details of the manner in which they cause infection in their respective hosts, but in general this process is much the same throughout the two genera.

Infection of the insect by the fungi via the alimentary tract probably occurs only rarely, if at all. The usual route of infection is by the penetration of the integument, especially the thinner intersegmental areas of the body wall and the appendages. Soon after the fungous spore comes in contact with the integument it begins to germinate, sending out a conidial hyphae that penetrates into the body cavity of the host where it develops rapidly at the expense of the softer tissues. Instead of producing a profusely branched mycelium, the hyphae usually become segmented and break down into their component cells, which are called "hyphal bodies." These short thick fragments of varied size and shape, containing a highly concentrated fatty protoplasm, multiply by budding or by fission until the insect's body cavity is filled with them. In some cases this fragmentation of the hyphae does not occur until later in the infection, after the mycelium is fairly well developed. The hyphal bodies may be very irregular in size and shape, or they may possess great regularity in this regard. The latter is frequently the case with hyphal bodies produced about the time of the host's death and just preceding spore formation.

After the hyphal bodies have been produced, if conditions of temperature and moisture are not favorable, instead of completing its development, the fungus may form chlamydospores. These spores are formed by each hyphal body which develops about itself a single wall of variable thickness depending upon the duration of this resting stage. By means of the chlamydospores the vitality of the fungus may be maintained through long periods of dormancy until the proper conditions for further growth are presented. The period from first infection to the formation of hyphal bodies or chlamydospores varies according to the host from 2 to 12 days. In general, the smaller the insect the shorter the period.

In the presence of sufficient moisture and adequately high temperature, the hyphal bodies and chlamydospores "germinate" with great rapidity. The hyphae thus produced may grow directly to the outer air and then produce a single conidium or set of conidia, according to the type of conidiophores. Sometimes, particularly when conditions of growth have been very favorable, a single primary hyphae may branch indefinitely, each ultimate branch becoming a conidiophore. As a rule, the number of germinating hyphae that develop from a single hyphal body does not exceed one or two, although instances occur in which the number is considerable. In the latter case the hyphae may branch and anastomose, forming a coherent mass known as a "stroma."

As a result of the germination of the hyphal bodies either sexual or asexual resting spores (zygospores or azygospores), or conidiophores bearing conidia are produced. In the case of the latter, as explained by Thaxter, the hyphae, arising directly or indirectly from the hyphal bodies, grow rapidly outward and burst through the less resistant portions of the hosts integument in spongy masses. In most instances the color of these masses is white, although the hues may vary from bluish white to pale or bright green or dull olive. There may be considerable variation in their general appearance, depending upon the species of fungus or the conditions of their development. In some cases the masses project barely beyond the body wall of the host and are confined to the points of emergence. These points are usually the thin intersegmental membranes through which the fungus projects in cushionlike rings, usually formed by simple conidiophores producing few or no branches outside the host's body. Each of these conidiophores gives rise to a single conidium. In other cases the external growth may be more extended, and the masses may coalesce so as to cover the entire body with a continuous layer of conidiophores which may form a mass several times as large as the insect itself. The external parts of these conidiophores are considerably branched, and the ultimate divisions of each conidiophore are arranged in a corymbose or digitate fashion. It is this occurrence of simple and compound conidiophores in different species which has led to the establishment of the two genera *Empusa* and *Entomophthora*. The difficulty in maintaining this generic separation is that the distinction is not absolute, and intermediate forms occur (e.g. in *E. culicis* Bra. and *E. apiculatus* Thax.). Compound conidiophores are sometimes found in species usually having the simple type, and simple conidiophores are commonly found in species having the compound type. In either case the growth of the conidiophores under optimum conditions of temperature and moisture takes place very rapidly and may give rise to the characteristic white masses in a few hours. Soon after tne masses of conidiophores appear, the production of conidia begins.

Thaxter explains the formation, discharge, and germination of the conidia as follows: The terminal portion of the conidiophore is termed the "basidium" and is usually swollen to a greater or less extent. From the apex of this basidium, the conidium commences its formation by the process of budding. The bud, or mother cell, increases until it reaches the normal size and shape of the conidium; then it becomes separate from the basidium by a cross partition. Within the mother cell thus formed is developed a single conidial spore. When the conidium is fully developed, the contents of the spore, as well as that of the basidium, begin to expand through the absorption of water. At first the contents of the basidium exert the greater of the two forces thus produced, and the columella is forced outward into the conidium toward which its convexity is thus turned. When the basidia are large and strong, this process may continue until the discharge of the conidium. Usually the contents of the conidium, being more dense than that of the basidium, finally exert a greater pressure and force the columella back into the basidium, thus reversing its former position. The sum of these opposing forces is very great, and as a result of their action a rupture of the wall takes place at the point where they are opposed—in a circle around the base of the mother cell. As a result of this rupture, the conidium is discharged violently into the air, often to a considerable distance. The columella commonly remains unbroken by this discharge, although it may be greatly stretched and hang down from the basidium as a tonguelike projection. In some instances, however, it may be broken or remain connected to the discharged conidium.

The conidia of the different species of the tribe Entomophthoreae vary in size and form; in fact, marked variation is sometimes apparent in the conidia of the same species. In size they range from about 10 to 75 microns in length. With regard to shape, the conidia vary from being spherical to being slender and tapering. They are usually hyaline, rarely slightly colored, and contain fine granular protoplasm or, more commonly, coarsely granular protoplasm with large fat globules. These fat globules may be so regular in size and shape that the conidia resemble asci filled with spores. The walls of the conidium are smooth, without spines or similar modifications, and possess an adhesive quality that serves to attach them to any object with which they come in contact. The basal portion of the spore is more or less papillate.

When discharged, the conidium, if it comes in contact with a suitable host, adheres to it and sends out a germination hypha, which enters its body in the manner already described. If the conidium lands in water, it gives rise to one or more hyphae that branch and elongate, growing constantly more attenuated, their protoplasmic contents becoming separated by successive cross partitions from the empty hyphae left behind. This growth may continue until the protoplasm is spent. The separation by cross partitions is common in the general growth of the fungus.

The process of germination is extremely variable but rarely lasts longer than a week and is usually much shorter than this. Germination ordinarily takes place soon after the conidium is discharged. If the discharged conidium falls neither on a proper host nor upon a wet surface, it proceeds to form what are called, "secondary conidia," a process that provides for further dissemination if the primary spore has fallen upon a substance unsuited for its proper development. The most common method of formation, according to Thaxter, consists in the production of a hypha of variable length which, growing vertically upward, becomes swollen at its extremity into a basidium and produces a conidium usually similar to that from which it originated. This secondary conidium is discharged in the usual fashion. It, in turn, may produce tertiary conidia, and the process continues until its vitality is exhausted or until it has come in contact with a suitable host. With some species, under unfavorable moisture conditions, the appearance of the secondary conidia may differ considerably from that of the primary conidium. They are usually almond-shaped, have thick walls, and are not discharged.

Occasionally, upon the examination of a conidiophore mass, a simple hypha is seen exceeding the conidiophores in size and projecting beyond them, often to a considerable distance. Some of these hyphae, called "cystidia," are so large that they may be readily seen with the naked eye;

others are not much different from the ordinary conidiophores. The function of the cystidia is not known unless, as Thaxter (1888) suggests, they are rhizoids or hyphae of attachment but functionless because of their position. The functioning rhizoids are hyphae that attach themselves to the substratum upon which the host rests and serve to hold the fungus firmly in position. The rhizoids may be simple or branched, and their termination may be modified into an expanded suckerlike structure of attachment. Rhizoids appear to be confined to certain species and usually accompany the digitate type of conidiophores.

*Formation of Resting Spores.* Within the body of the insect; infected with an Entomophthoreae, a process or phenomenon frequently occurs by which special spores are formed that are highly resistant to conditions that would ordinarily destroy the conidia. These "resting spores," as they are called, may be formed by an asexual process, in which case they are known as "azygospores," or they may be formed by sexual union and then are known as "zygospores." The resting spores are usually spherical, of large size, contain highly refractive fatty contents, and are surrounded by triple walls. The outer wall of the spore is thin and represents the wall of the mother cell, the second is thicker, and the innermost wall is usually as thick as or thicker than the other two combined.

Azygospores may be formed in a variety of ways. The simplest process is that by which the contents of a hyphal body become directly converted into a resting spore. Or azygospores may be formed from hyphae of germination arising from chalmydospores or hyphal bodies, or by direct lateral budding from them. Sometimes azygospores may be produced interstitially—between fungous cells—and this frequently results in spores having very irregular shapes.

Zygospores are also formed in a variety of ways, although sometimes the sexual nature of the spore is not well marked, in which case this may represent a transitional form from the truly sexual to the entirely asexual processes. The method of true zygospore formation is, as might be expected, by conjugation of two different hyphal outgrowths; these meet, the intervening walls are absorbed, and the contents of the two mingle. In many cases a bud then appears on one or both of the gametes, increases rapidly, and becomes the zygospore. Sometimes the spore develops as a terminal swelling from the end of one of the conjugating hyphae. Other methods of formation have been observed.

*Hosts and Habitats of Entomogenous Entomophthoraceae*. Any attempt to summarize the insect hosts of a particular group of microorganisms must take into account the fact that the host distribution is to some extent dependent upon the relative amount of study made of the groups concerned. In the case of the insect hosts of Entomophthoraceae, the Diptera appear to be the order of insects the members of which are found to be the most frequently infected, and from which the greatest number of species of this group of fungi have been isolated. The Hemiptera are the next greatest sufferers, followed by the Lepidoptera and Coleoptera. Certain species of Orthoptera, Hymenoptera, and Neuroptera are also known to be susceptible to these fungi. The adult stage of the host insect is affected more frequently than either the larvae or pupal stage. In insects with incomplete metamorphosis the nymph is almost as susceptible as the adult.

Although to a degree there is some specificity of hosts for each species of the tribe Entomophthoreae, this is by no means certain or uniform. Some species of Entomophthoraceae infect a wide range of hosts, including insects in indifferent orders; others have been found only on a single insect species or on a closely related group of insects. No infrequently two different species of fungi are found on one species of host or even on a single host.

Entomogenous Entomophthoraceae are found in a variety of habitats, but usually they develop best in areas where there is a constant and rather abundant supply of moisture. The edges of ponds and brooks in shaded places are a particularly suitable environment for many of them. Some species do well in drier situations as long as periods of occasional moisture occur, enabling conidia to be produced. Damp foggy weather frequently makes the fungi more conspicuous, since the moisture causes them to become distended.

Just before dying from the infection, many hosts seek elevated positions, crawling upward on blades of grass or other food plants. This incidentally, makes it possible for the conidia to be discharged over a considerable area. Other favorite positions assumed by the infected insects are on the undersides of leaves in shady areas or about houses.

*Cultivation of Entomogenous Entomophthoraceae*. Only a few species of entomogenous Entomophthoraceae have been cultivated apart from this insect hosts. The fact that success in this regard has been obtained with a few species indicates that probably most of the remaining species can be cultivated if the right combination of substratum and growing conditions can be found.

The first to obtain the complete life cycle of one of these fungi in artificial culture apparently was Speare (1912), who succeeded in thus cultivating *Entomophthora pseudococci* Speare, a parasite of mealybugs in Hawaii. As media he used potato, potato agar, oat agar, and radish, the latter two being found the most suitable. In 1929, Sawyer reported the cultivation of *Entomophthora sphaerosperma* Fres. and an unidentified species of *Empusa*, the most satisfactory media in this instance being potato, swordfish, and pork.

Sawyer's studies revealed several interesting features connected with the cultivation of the two species of entomophthoraceous fungi with which he worked. For one thing, liquid nutrient media favored luxuriant mycedial growth, whereas solid media favored the production of hyphal bodies and reproductive phases. Carbohydrates and fats did not appear to be essential to the growth of these fungi, but the substratum had to contain protein that are quickly liquefied by the proteolytic enzymes secreted by the fungi. The hydrogen-ion concentration of the media was also important, with pH values slightly on the acid side of neutrality being the optimum for growth and development of the organism. Atmospheric humidity did not materially influence growth or reproduction, although too much moisture in the substratum inhibited the latter. Conidia did not germinate below a relative humidity of about 70 per cent at 21°C. This temperature (21°C.) appeared to be the most favorable for growth and reproduction; the maximum temperature was 34°C. The exact minimum was not determined, but Sawyer found that the conidia of both species with which he worked could be frozen for several days and still germinate upon return to room temperature. Vegetative growth and the production and germination of conidia took place in total darkness as well as in light. Best growth was obtained when large quantities of inoculum were transferred to the media used. Cultures held at room temperature were best transferred once every 10 days.

The artificial propagation of entomogenous Entomophthoraceae may also be accomplished by infecting fresh hosts. Thaxter (1888) developed a method of doing this by using a tightly covered jelly tumbler in which the upper portion is separated from the lower by a round piece of wire netting. By placing the hosts to be infected is the lower of the two chambers and fastening a specimen carrying the fungus in the upper one,

Thaxter found that the living hosts below acquired the discharged spores through the netting and thus became infected.

**Examples of Empusa and Entomophthora Infections**

It is not practical for us to attempt here a discussion of all the species of entomogenous Entomophthoraceae and the infections they cause. The essential relationships involved may be exemplified, however, by a brief consideration of a few of the better known species.

Since the systematics of this group of fungi is still unsettled, we shall not be too concerned about the eventual validity of the generic names to be used in certain cases. Inasmuch as most modern mycologists use the name *Empusa* to include species of Entomophthoreae having simple or unbranched conidiophores and multinucleate conidia, and the name *Entomophthora* for species with digitate or branched conidiophores and uninucleate conidia, we shall do the same. Nowakowski's genus *Lamia* is not now generally recognized, and we shall not use it here.

*Empusa muscae Cohn*. The first description of this fungus, commonly found infecting the housefly, *Musca domestica* Linn., was that given by DeGeer in 1782. About a century ago Cohn (1855) gave it the name *Empusa muscae*, and it stands today as the type species of the genus. A considerable number of Diptera, in addition to the housefly, are subject to attack by this fungus, the most common of which are reported to be species of the genera *Lucilia* and *Calliphora*. Syrphidae are also susceptible. Distribution is apparently world-wide; most of the significant observations on the fungus have been made in the United States, Europe, and South America.

Infected flies are usually found indoors; they attach themselves to the walls and ceilings of houses and other buildings in the lifelike position. Close inspection of flies killed with the fungus usually reveals on the wall or windowpane a distinct halo of discharged spore encircling the insect.

The life cycle of *Empusa muscae* may be distinguished by its relatively large pointed bell-shaped spores each of which usually contains a single large oil globule. The pellicle of protoplasm (from the basidium) that surrounds the spores when discharged one such a surface as that of a glass slide give them a characteristic appearance. (Fig. 1.1).

*Empusa grylli* (Fres.). Fresenius (1858) originally described this fungus from a ground beetle under the name *Entomophthora Grylli*; but

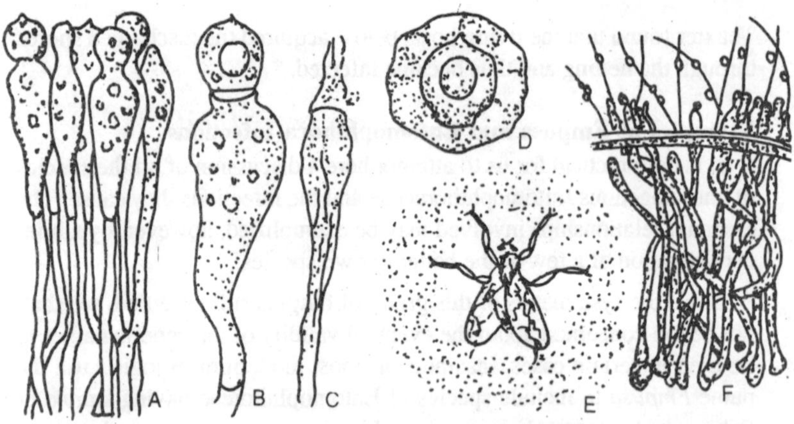


**Fig. 1.1.** *Empusa muscae* Cohn, the fungus commonly found infecting the house-fly and other flies. *A*. A group of conidiophores showing conidia in several stages of development. *B*. Basidium bearing conidium before discharge. *C*. Basidium after discharge of conidium. *D*. Conidial spore discharged upon a glass slide and surrounded by a mass of protoplasm from the basidium. *E*. A fly infected with *E.musae*, the ejected spores forming an aureole about the dead insect. *F*. Mycelium and conidiophores penetrating cuticle of insect (diagrammatic).

since the conidiophores are of the simple type, the generic name *Empusa* is probably more applicable, and it has been recognized as such by most subsequent taxonomists. The conidia of this fungus are ovoid to pear-shaped, the conidiophores may coalesce externally when growing luxuriantly, cystidia are wanting, and the resting spores are spherical and coloreless. Its hosts include numerous species of Lepidoptera and Orthoptera, especially short-horned grasshoppers. Its prevalence on grasshoppers or locusts accounts for the designation "common locust fungus," often applied to it.

One of the first reports in American literature having reference to the destruction of locusts by a fungous infection is that of Bruner (1883) in which the author states that numerous instances of "internal fungoid growths" had come to his attention during the preceding 12 years. In 1896, in South Africa, grasshoppers had been found dying from a fungous disease. Some of the diseased specimens were sent to the United States, and a fungus was isolated. This unidentified fungus was distributed to planters, who in many cases reported considerable destruction of grasshoppers by the use of the microorganism. The possibilities attending the use of fungi in the control of grasshoppers appeared promising enough to initiate further studies, which were promoted under the direction of L. O. Howard of the Bureau of Entomology of the U.S.D.A. In 1902 Howard summarized the various reports on the effectiveness of fungi as a control agent and concluded that the results obtained did "not justify very sanguine hopes." He also pointed out that some of the cultures being distributed by various agencies and laboratories were not *Empusa grylli*. Some of them were saprophytic contaminants, and others were species of *Mucor* that did seem to have some insect-killing properties. A record was also made of the susceptibility of grasshoppers to the so-called "chinchbug fungus," *Beauveria globulifera* (Speg.). In the years following Howard's report, the artificial use of *Empusa grylli*, as well as other fungi, as an agent of insect control, lost favor. Its destructiveness in natural outbreaks was recognized, but its effectiveness was too dependent upon optimum conditions of temperature and moisture to be a practical means of artificial control.

The symptomatology of an *Empusa grylli* infection in grasshoppers is fairly characteristic. Skaife (1925), in South Africa, has described the disease somewhat as follows: The dying insects climb as high as they can on the grass stems and on the twigs of bushes with their heads pointing upward. Just before death, which usually occurs about 5 or 6 days after infection takes place, they losen their hold with their claws and embrace the stems with their legs. After death, which occurs while they are in these elevated positions, their legs stiffen and the dead bodies remain hanging in this position for several days until they are finally blown away by the wind or washed down by the rain. Skaife observed that the majority of the insects infected with the fungus die in the late afternoon, usually between 3 and 7 PM. An hour or so after death a fine furry or velvety growth appears growing from the intersegmental membranes, from the joints of the legs, around the neck, and at the base of the antennae. This growth usually has white, buff, or greenish color and it consists of innumerable club-shaped conidiophores that project from the insect's integument. At the end of each conidiophore a conidium is produced. These conidia are discharged generally in the evening when the live grasshoppers are clustered together for the night. This, of course, facilitates the transmission of the fungus from the diseased insect to a healthy one. According to Skaife, there are several different strains of *Emupsa grylli* varying in virulence, and apparently individual grasshoppers vary in their susceptibility to any particular strain. About 1 per cent of the dead grasshoppers fail to produce the external growth of conidiophores. When these insects are opened they are usually found to be filled with the thickwalled, resistant resting spores.

In South Africa, Skaife found that the disease develops only in those localities which had a rainfall of over 14.5 inches during the 6-month period. The disease did not appear in four areas that had received over 4 inches of rain in only 1 month. A rainfall of 14 to 15 inches over a period of 3 months, however, appears to be sufficient to start the disease in South Africa. In the United States it is generally agreed that warm humid or wet weather is necessary to enable the infection to develop. During dry weather there is practically no extension of the disease. It is unfortunate that grasshoppers do their greatest damage in the dry seasons, when conditions least favorable for the development of the fungus prevail.

*Entomophthora sphaerosperma* Fres. One of the best studied entomophthorous fungi is *Entomophthora sphaerosperma*, a fungus first recorded as a parasite of the cabbage butterfly of Europe, *Pieris brassicae* (Linn.), by Fresenius (1858), who briefly described it and gave it its present name. The fungus has several times been known by other names (*Tarichium sphaerosperma* Cohn, *Empusa radicans* Brefeld, *Entomophthora radicans* Brefeld, *Entomophthora phytonomi* Arthur, and *Empusa sphaerosperma* (Thaxter), but the name and combination originally used by Fresnius have retained the validity. It was first reported in the United States by Arthur (1886)—under the name *Entomophthora phytonomi*—as a parasite of the clover leaf weevil, *Hypera punctata* (Fabr.). Since then it has been found on a wide variety of hosts in many parts of the world. In fact, its host range includes nearly every order of insects, the most common being Diptera, Hemiptera, and Lepidoptera. In numerous instances insects are subject to widespread outbreaks of infection by this fungus, which thus is frequently of considerable economic importance in the natural control of certain insect pests.

The morphology and development of *Entomophthora sphaerosperma* has been described by several authors, but one of the most complete accounts in that by Sawyer (1931). This worker succeeded in growing the fungus on swordfish and potato. From such cultures, which grew best at temperatures of from 18 to 21° C., Sawyer was able to make a careful comparison of the various stages in the life cycle of the fungus. The conidia are narrowly elliptical, with a rounded apex and a tempering base, and have an average size of 7 by 22 microns. A single membrane encloses the spore, and the apex is frequently covered by a detachable, transparent gelatinous cap. The protoplasm is finely granular and non-vacuolate in the newly formed spore, although vacuoles begin to form as soon as the spore has a single centrally located nucleus. The mature conidiophores are digitately branched at the distal end. A conidium develops at the end of each branch, a thin wall separating it from the conidiophore, which has also formed a cross wall at this point. After a given interval the spore is violently discharged. If the spore lands in a suitable environment germination begins with a slight outward bulging of the conidium at any point on its surface. As growth proceeds, a tube is formed at the tip of which may arise a secondary conidium; or, if it is able to penetrate a susceptible host, the germ tubes may continue to develop, forming a mycelium and hyphal bodies. Rhizoids may also eventually be formed. These structures usually arise from the hyphae in the thoracic region of the insect and attach the host firmly to the substratum. The rhizoids constitute the only structure of the fungus that is not developed on artificial culture media as used by Sawyer. Resting spores are formed asexually (azygospores), their formation being subject to such factors as temperature and the nature of the artificial media upon which the fungus is being grown.

Although *Entomophthora sphacrosperma* is distributed widely and on many different host species, detailed studies of epizootics caused by this fungus have been made in relatively few instances. One such study was accomplished in South Africa by Ullyett and Schonken (1940) in the case of such an epizootic among larvae of the diamondback moth, *Plutella maculipennis* Curtis, which are first miners and later surface feeders on cruciferous plants. These workers observed the sporadic occurrence of the disease in fields in Transvaal where, at times it was an important factor in the temporary control of the insect.

During the early stages of the infection the *Plutella* larva shows no significant external symptoms. As the disease becomes more advanced, however, the insect becomes restless and changes in color from green to a yellowish green. The yellow color becomes intensified, and the larva becomes increasingly sluggish in its activity as the disease progresses. With the approach of death, the body becomes somewhat distended; and immediately after death and before the aerial hyphae appear the larva is extended on the leaf and is turgid and brittle. Upon being lifted, the body breaks easily, but there is little or no liberation of fluids. This last feature is a convenient one for making field diagnoses. Early larval instars appear to be more resistant to infection than do the later ones, fully grown larvae being very susceptible. The development of the fungus in the body of the larva is essentially the same as has been described as characteristic of the group. Germination of the conidia requires prior contact of the latter with free moisture. Thus it is that epizootics of the disease break out only after heavy rains. The germination hypha penetrates the integument of the host and enters its body cavity where growth rapidly follows, giving rise to a branching mycelium. Infection of the *Plutella* lara almost invariably takes place at the anal extremity, the hyphal threads running forward longitudinally in the body cavity. This forward growth of the mycelium apparently is a normal attribute of the fungus, since regardless of where experimental infection takes place the initial growth is toward the cephalic end of the insect, taking place against the flow of the blood stream. The body cavity gradually becomes filled with the mycelium. Just before the death of the host the hyphae break up into short lengths, and immediately after death the mycelium is a compact mass of short stout lengths of hyphae (the hyphal bodies) which completely fill the body of the insect. The reproductive phase of the fungus begins with the death of the insect. Rhizoids anchor the dead larva to the leaf or other subtratum, and hyphal branches emerge and ramify over the surface becoming interwoven at right angles to one another until the body is covered with a grayish-white felt of fructifying mycelium. Branched conidiophores arise from the thallus thus formed and bear the characteristic elongate conidia. When the mature conidia are discharged they come to lie on the leaf surface immediately surrounding the dead insect. It is from this source that healthy larvae pickup the spores as they crawl over them. At the completion of the aerial fruiting of the fungus, the center of the thallus turns brown in color. Within the body of the dead host azygospores are formed from the hyphal bodies. These spores remain within the mummified host until it becomes broken up or until favorable weather conditions induce germination. Under dry conditions the viability of the azygospores is retained for long periods of time.

The effect of epizootics caused by *Entomophthora sphaerosperma* on the *Plutella maculipennis* population was examined by Ullyett and Schonken (1940) from both the theoretical and the practical standpoint. In brief, however, these workers concluded that the fungus, although it produced a decided immediate reduction in the host population during the favorable weather conditions under which it acted, was ultimately responsible for an increase in the average density of the host and, through this, for the occurrence of economic damage to the crop. The principal reason for this situation appeared to be the concurrent destructive effect on the insect parasites and predators that normally held the population density low enough to avoid serious crop damage. Thus the intervention of the disease in a existing equilibrium system resulted in the destruction of permanent mortality factors (insect parasites and predators) and their replacement by a temporary mortality factor (the fungus). When the activity of the fungus ended, the host population was able to recover more rapidly than its normal parasites and thus to attain a higher density level than before. This is an excellent example of the fact that occasionally a destructive disease may actually be a detriment rather than an aid in the long-range control of an insect pest.

An instance in which the artificial distribution of *Entomophthora sphaerosperma* affected the localized control of a destructive insect is provided by the work of Dustan (1924), who used it against the European apple sucker, *Psyllia mali* Schmidb., in Nova Scotia. This host was first found infected in considerable numbers in 1920 and again in 1921, and in the humid years of 1922 and 1923 artificial distribution of the fungus practically exterminated the insect in certain orchards. The fungus was spread by pinning leaves bearing diseased insects to leaves in the orchard being treated and by the liberation into the orchards of infected adults still capable of flying about. About a week after such "planting," the fungus began to appear on the apple suckers in the orchards being treated. It spread very rapidly and was generally distributed throughout the orchards within 2 or 3 weeks if weather conditions were favorable. Dustan was able to initiate outbreaks of the diseases earlier than would naturally be the case by building up active epizootics in caged material early in the season and the introducing this material into the orchards.

*Entomophthora aulicae* (Reich.). The so-called "brown-tail fungus," *Entomophthora aulicae*, attacks the caterpillars of the brown-tail moth, *Nygmia phaeorrhoea* (Donov.), as well as more than a dozen other insect hosts. Thaxter (1888) considered this fungus as a strain of *Empusa grylli* (Fres.), but it has since been recognized as a distinct species. It has long been known in Europe and was first recorded in the United States by Thaxter is 1888. Both natural and artificial epizootics of the fungus on the brown-tail caterpillar have been reported.

In the spring and early summer the caterpillars leave their nests to feed on young leaves, and in the autumn they again leave before the webs of the new brood are closed for the winter. During both these periods, Speare and Colley (1912) found that the fungus could be effectively used against the insect, depending on the presence of favorable weather conditions, which appear to consist of warm nights and damp atmosphere. The natural epizootics also occur during these periods and under essentially the same conditions.

After developing methods for the artificial propagation of *Entomophthora aulicae* on brown-tail caterpillars in cages, Speare and Colley came to the following conclusions with regard to their attempts to use the fungus against this insect in the field. They were able to effect the destruction of caterpillars in great numbers and over considerable areas, although the general effectiveness of this method was variable since the factors of temperature and humidity could not be controlled. Nevertheless the introduced disease could usually be depended on to kill from 60 to 100 per cent of the caterpillars in the areas where diseased individuals were distributed. Best results were obtained in localities where the disease was not known to occur naturally. The fungus usually lives over winter from the autumn infection and reappears early in the spring, reducing the caterpillar population. For this reason the autumn infection is doubly effective. Artificial distribution appears to be most easily accomplished in sprout woodlands and in pastures where ordinary methods of control (e.g., spraying and cutting) are not employed and where the caterpillar nests are more readily accessible.

Speare and Colley's (1912) report also contains a description of experiments with an unidentified *Entomophthora* on larvae of the gypsy moth, *Porthetria dispar* (Linn.). They concluded that this entomophthorous disease was not a promising one for artificial use.

*Entomophthora fumosa Speare.* It has long been recognized that in Florida the citrus mealybug, *Pseudococcus citri* Risso, is a pest of only secondary importance. In certain parts of California, on the other hand, it is considered one of the most serious enemies of citrus plants. The reason for this interesting difference was pointed out by Speare, who in 1922 published a revealing account on the natural control of the citrus mealybugs in Florida. According to Speare, the principal reason that *Pseudococcus citri* Risso is not an economically important pest in Florida is the fact that it is held in check by an entomogenous fungus, *Entomophthora fumosa*. The climatic conditions in Florida during the growing season are optimum for the development of the fungus, whereas this is not the case in California. In Florida the warm temperatures and the rainy season coincide; in California they do not—the rainy season occurs during the cooler winter months. Thus it is that in Florida the fungus has an opportunity to flourish and to bring about the natural destruction of its mealybug host.

Although Speare originally gave the fungus the generic name *Entomophthora*, it would probably be more appropriate for it to bear the name *Empusa*, since the conidiophores are of the simple type. This designation depends, of course, upon the importance one is inclined to give the type of conidiophore in the generic separation of the Entomophthoraceae. The conidia of the fungus are more or less fusiform in shape, approximately 9 by 18 microns in size, and distinctly smoke-colored. Resting spores are formed and are usually spherical, black, and provided with a hyaline protuberance or appendage.

The symptoms of the disease caused by *E. fumosa* have been described by Speare (1922). One of the earliest signs of infection may be elicited by the use of a penknife or other bluntly pointed instrument which, when pressed against the body of a healthy mealybug, creates a depression that quickly returns to its original position when the pressure is withdrawn, much as a rubber ball would do. In mealybugs in the early stages of the disease, however, the body wall ruptures very easily when pressed with a penknife, allowing a droplet of milky fluid to exude. This occurs in the infected insects, even though to all outward appearance they may seem entirely healthy and active (at times their movements may be somewhat sluggish). If similar treatment is applied to a recently dead mealybug, which is also lifelike in appearance, the depressed area remains sunken. After the insect has been dead from 12 to 18 hours, its body is more or less solid to the touch, and if considerable pressure is applied to the penknife, the body will cut like a piece of cheese. No significant change in the external appearance of the mealybug takes place until about 24 hours after the first symptoms have been observed. At this time, and later, the infected insects may present one of two types of appearance. Most of the dead insects will appear to be enveloped in a dark slate-gray woolly covering; which consists of the fungal conidiophores and the developing conidia. A few of the insects, on the other hand, may appear jet black in color, sometimes almost glistening, with the body surface of the insect smooth instead of woolly. The jet-black color is due to the large numbers of spherical black resting spores formed within the body of the insect. The color of the black spores is transmitted through the thin translucent body wall still intact. In both of the two types of infection the mealybugs are attached rather lightly to the host plant by the insertion of their proboscises.

In orchards of grapefruit, Speare observed an extremely rapid spread of the fungous infection among the mealybugs during the months of June, July, and early August. In one grove, the percentage of mortality jumped from 18 to 64, an increase of 46 per cent, in but a week's time (June 22 to June 29). The role of the fungus in the natural control of mealybugs is also emphasized, in Speare's opinion, by the fact that following the application of fungicides, which destroy the entomogenous fungus as well as that of the citrus scab, the numbers of mealybugs rose sharply in the groves so treated.

Incidentally, certain Fungi Imperfecti have from time to time been reported as being parasitic on mealybugs. Species of *Aspergillus* and *Cephalosporium* have been mentioned in this connection. Boyce and Fawcett (1947), for example, relate that during the propagation of mealybugs in California insectaries, a fungus closely related to *Aspergillus parasiticus* Speare proved to be a potentially serious parasite under conditions of high humidity and moderate to high temperatures. Efficient ventilation of the insectaries, careful irrigation of the host plants, and a lowering of the temperature to 20°C., or below, are suggested control measures. A fungus that has been named *Endosclerotium pseudococcia* Har. & McK., occasionally kills large numbers of mealybugs, including the Comstock mealybug, *Pseudococcus comstocki* Kuw., in the apple orchards of Virginia and other eastern states. When conditions for growth are unfavorable, the fungus produces a highly resistant compound sclerotium which may remain viable several months.

## INFECTION CAUSED BY MASSOSPORA

Of the remaining genus of Entomophthoraceae, *Massospora*, the only well-studied entomogenous species is *Massospora cicadina* Pk., named and described by Peck in 1879, although apparently seen by Leidy as early as 1850. It is a parasite of the seventeen-years locust, or periodical cicada, *Magicicada septendecim* (Linn.), and had been observed particularly in the Eastern and Middle Western parts of the United States.

The occurrence of the fungus was subsequently recorded in the reports of a number of entomologists, but it remained for Speare (1921) to give the first adequate description of the disease itself, as well as that of the microscopic characters of the fungus. He also showed the relationship of *Massospora cicadina* to the other entomogenous Entomophthorales. Goldstein's (1929) cytological study of the fungus furnished additional information on this very interesting parasite.

In some outbreaks at least, the fungus appears to be confined largely, although not exclusively, to male insects, and in the resting spore condition it usually parasitizes spent individuals, females as well as males having been found in this condition. It confines its vegetative growth to the softer tissues in the posterior segments of the cicada's body. Most of these tissues are completely destroyed, and as a result of the destruction of the flexible intersegmental membranes, the posterior abdominal segments, beginning with the last segment, slough off. The insect, however, remains alive for a considerable period, continuing to fly and crawl about. This unusual phenomenon is so striking that it is readily noticed by all observers.

One of the outstanding characteristics of *Massospora cicadina* is that, unlike species of *Empusa* and *Entomophthora*, it produces conidia within the body of its host rather than on its surface, where they are violently discharged from the conidiophores. Within the abdomen of the cicada, the conidia cohere with one another forming clumps, clusters, or a mass that is exposed when the insect's abdonminal segments drop off. The movements of the insect then aid in scattering the conidia of the fungus. The conidia are oval in form, about 12 by 15 microns in size, and have a distinct but not prominent papilla. Unlike those of most other Entomophthorales, the conidial walls are regularly verrucose. Resting spores, found in certain individuals, are slightly brownish spherical bodies with reticulations that give them an appearance similar to the design on golf balls. Attempts to cultivate the fungus on artificial media have not been successful.

Just how or where the fungus survives the 16¾ years of the host's immature and subterranean existence is not known. It is possible that infection takes place while the insect is underground, or the fungus may exist on species of biennial cicadas when its regular host is absent. Its importance in the natural control of the periodical cicada is probably not great, if Speare's observation that the infection is confined largely to spent males is correct. Goldstein (1929), however, found the fungus in both males and females, and most of her specimens containing resting spores were females whose bodies still contained many eggs. Such a situation would enhance the importance of the fungus from an economic viewpoint.

## INFECTIONS CAUSED BY CHYTRIDIACEOUS AND BLASTOCLADIACEOUS FUNGI

Among the more primitive of the phycomycetous fungi are those belonging to the order Chytridiales and Blastocladiales. The "chytrids," as the former are commonly called, occur mainly in fresh water on a variety of substrata. A number of species live in the submerged and empty exuviae, or castoff integuments, of the immature stages of certain insects.

Truly entomogenous chytridiaceous fungi, however, have been reported by a number of mycologists. Wize (1904) found such a fungus in Coleoptera (*Cleonus, Anisoplia)* larvae and pupae collected in the Ukraine, and Sparrow (1939) reported on similar observations made on dipterous pupae collected in the United States by Thaxter. In both instances the fungus appears to have been the same, *Myiophagus ucrainicus* (Wize) [*Myrophagus*]. The body contents of the diseased insects are almost completely disintegrated and replaced by an orange to reddish mass of fungous material. What is apparently this same fungus has also been found in dipterous pupae in England, and in Bermuda, Canada, and the United States in scale insects. Its ocurrence in scale insects (purple scale, chaff scale, oyster-shell scale, red scale, long scale, and soft brown scale) is particularly interesting and the chytridiosis produced probably deserves study from the standpoint of biological control. Purple scales have been infected by spraying them with aqueous suspensions of the fungus, and mealybugs have been found experimentally susceptible.

The entomogenous Blastocladiales are better known than are the apparently few chytrids that parasitize insects. Nevertheless these are confined largely to one group, the family Coelomomycetaceae, which parasitizes mainly mosquito larvae.

### Coelomomyces Infections

The first recognized infection caused by a member of this group of fungi was that discovered by Keilin in 1921, occurring in larvae of the mosquito *Aëdes albopictus* (Skuse) (= *Stegomyia sculellaris* Walker) collected in the Federated Malay States. To this fungus, which he believed to be related to the Chytridineae, Keilin gave the name *Coelomomyces stegomyiae*. The next year Bogoyavlensky (1922) described, under the name of *Zografia notonectae*, a fungus parasitic in the body cavity of *Notoceta* (Hemptera) collected from ponds in Moscow. Keilin (1927), however, showed this organism to be of a nature similar to that of the fungus he studied, thus giving it the name *Coelomomyces notonectae* Additional species have been isolated from mosquitoes (*Anopheles, Culex, Psorophora*, and *Uranotaenia*) in other parts of the world, including 2 species from India, 1 species from Africa, and 11 species and 2 varieties from the United States.

**The infection.** The manner in which mosquito larvae become infected with *Coelomomyces* has not been determined with certainty; i.e., it is not clear whether invasion takes place via the digestive tract following the ingestion of the zoospores or the sporangia, or directly through the insect's integument. In any case the principal development of the fungus occurs within the body cavity of its host. Depending upon how far the infection has progressed, only certain regions of the body, such as the gills or the posterior segments, may appear involved, or the entire hemocoele may appear filled with the parasite. Frequently the mycelia and spores of the fungus may occur even within the head of the larva. (Fig. 1.2).

The hemocoele of an infected larva in the fourth instar may appear as a solid mass of sporangia, which gives the larva an opaque aspect. The color of such larvae may be dull or yellowish-white, bright yellow, orange, or even a dull reddish-brown.

In most instances the fungus completes its development during the larval stage of the mosquito. Sometimes, however, when infection occurs late in larval life the fungus completes its development in or persists through the pupal and adult stages. In surviving adult females the infection is usually confined to the ovaries. In other cases it occurs throughout the body of the adult, causing its death. Infected pupae may be unable to emerge as adults; especially is this so once the sporangia have developed. It is probable that the majority of infections begin during the larva's first instar, since fully developed sporangia are found in the third and fourth larval instars. Since this period of larval development is usually one of only 6 to 10 days, it indicates the very rapid development of the fungus. Actually, under normal conditions in nature, it takes only 2 or 3 days for the thalli to develop into sporangia. According to Iyengar (1935), infection generally begins in the thoracic region of the larva and spreads posteriorly into the abdominal segments, travelling along the adipose tissue on which the fungus lives. As a result of the attack by the fungus, the fat body loses its characteristic appearance and shrinks, the nuclei of the fat cells disintegrate, and finally the fat body completely disappears. In place of the fat body, a thin membrane filled with many dark-brown

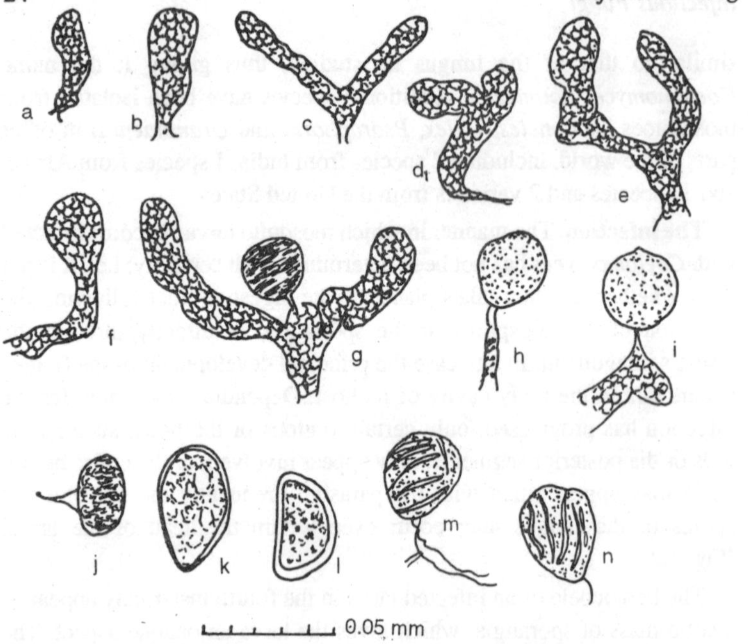


**Fig. 1.2.** *Coelomomyces anophelesica a* and *b*, young mycelia; *c-f*, older mycelia; *g* and *i*, formation of sporangium; *h* formation of sporangial wall and withdrawing of contents of mycelium; *j*, young sporangium with attached strand; *k*, vacuolated stage of sporangium; *l*, older sporangium with dense contents and thick wall; *m* and *n*, mature sporangia with partly developed ribs and connecting strand.

pigment granules remains. Only rarely does an infected larva of the fourth instar have any fat tissue. Some fat tissue may remain if the infection has been of a low intensity. One reason for an infected larva's rarely completing its metamorphosis into an adult mosquito is the suppression of the imaginal buds by the infection. In a fourth-instar infected larva, the wing and leg rudiments are underdeveloped as compared with their appearance in a healthy larva of the same instar.

No accurate survey has yet been made as to the over-all extent to which mosquito larvae are infected with *Coelomomyces* fungi. In some areas the incidence of infection appears to be very high, while in other areas infected specimens can be found only on rare occasions. Muspratt (1946) estimates the mortality, in pools that were under his observation in Rhodesia from 1941 to 1945, to be as high as 95 per cent of the larvae that hatched from eggs during the rainy season. Of those larvae which reached the fourth instar, at least 9 out 10 were infected and subsequently died. Offhand, at any rate, such data would make it appear that these fungi may possibly have some use in the biological control of mosquitoes, at least in the tropics. On the other hand, some observers have found the ratio of infected larvae to healthy ones to be so small in the areas they studied that the fungus did not appear to have much importance as a natural control agent. The biological control possibilities are under consideration by several groups of investigators as this is being written, and the outcome of these as well as other studies must be awaited before it will be safe to draw any definite conclusions in this regard.

*The Fungi.* As has already been mentioned, the *Coelomomyces* fungi are phycomycetes belonging to the family Coelomomycetaceae of the order Blastocladiales. It was at first believed that like those of most species of Blastocladiales, the resting bodies of *Coelomomyces* required a period of desiccation before being able to germinate. Muspratt goes further to suggest that in order to bring about the germination of the resting sporangia, it may be necessary to use rain water and allow it to evaporate in the sun to about one-third of its volume before infection can be expected, the germination perhaps being regulated by a slight increase in the concentration of the soil mineral salts in solution. In 1947, however, Couch and Dodge found germination to occur in the case of three species without the previous drying of the resting bodies. These authors nevertheless feel that the fungi belong to the order Blastocladiales, particularly since in structure and method of swimming the zoospores are typical of this order.

Germination of the sporangia was first witnessed by De Meillon and Muspratt (1943). These workers observed the process to be somewhat as follows: When about to germinate, a sporangium first loses its oil droplets and the interior becomes granular. A slight bulge then appears on one side. As the bulge enlarges, the outer hard shell ruptures and two thin internal membranes appear. The content of the sporangium flows out and is confined within the innermost of the thin membranes, both of which become increasingly more visible and more widely separated. The contained zoospores become less densely packed and accordingly begin to move about with increasing speed until the interior of the sporangium, and the extruded membranes, are a seething mass of zoospores. Within a few minutes thereafter the zoospores find their way out to the exterior until the sporangium and rent membranes are empty. This whole process may take a day or two; but once the zoospores flow into the extruded membranes, complete liberation takes place in a few minutes. The zoospores have single flagella about four times the length of the zoospores, which measure approximately 4 microns.

Further observations on the germination of the resting sporangia were reported in 1947 by Couch and Dodge. They also observed germination to begin with a swelling of the contents which causes the outer wall to split along a preformed line. The contents continue to swell, bulging out through the fissure to form a dome-shaped mass surrounded by the inner spore wall. The zoospores are formed 24 to 36 hours later, and the entire exposed part of the inner wall quickly swells and dissolves to set the zoospores free. The two thin membranes described by De Meillon and Muspratt are interpreted by Couch and Dodge as a single membrane that gelatinizes. Otherwise the observation of these two pairs of workers are essentially similar.

Within the body cavity of the insect, the fungus develops a coenocytic, aseptate mycelium without rhizoids or cell walls, and surrounded only by a plasma membrane. This last characteristic enables the fungus to absorb its food directly over its entire surface. The hyphae are irregularly or rarely dichotomously branched and consist of single threads of more or less uniform diameter except at certain points. The hyphae give rise to irregularly shaped hyphal segments that are severed from the parent hyphae by division or by the thinning down of the basal connection until it is broken away by the movements of the insect's body. The sporangia are formed as are the hyphal segments. They break away from the mycelium before the formation of a definite wall and complete their development within the hemocoele of the insect. The thick-walled sporangia fill the body cavity and are usually set free upon the disintegration of the insects body. The sporangium is surrounded by an exceedingly thin, hyaline, smooth, outer membrane derived from the old plasma membrane. The wall is two-layered, the outer layer usually being the thicker. It may be smooth pitted, banded, striated, ridged, or otherwise sculptured or ornamented. Dehiscing occurs by a preformed longitudinal slit. All these characteristics have been stated descriptively by Couch (1945), who revised the genus *Coelomomyces* Keilin and erected the family Coelomomycetaceae.

## ASCOMYCETE AND DEUTEROMYCETE (FUNGI IMPERFECTI) INFECTIONS

It is more of convenience than for any other reason that we place the ascomycete and deuteromycete infection under one general heading.

Most authorities recognize the majority of species now relegated to the class Deuteromycetes (Fungi Imperfecti) to be, in reality, species of Ascomycetes whose relation with an ascogenous stage has not been established. In other words, since most of the Deuteromycetes appear to be asexual (imperfect) stages of recognized or unrecognized species of Ascomycetes, it is expedient for us to consider the two groups together. Especially is this a convenient procedure in dealing with such entomogenous fungi as those which parasitize scale insects and whiteflies and which, in many cases, are known by different names in both their sexual and asexual (perfect and imperfect) stages.

As has been expressed by Wolf and Wolf (1947), the Ascomycetes "possess a multiplicity and complexity of architectural design and a seemingly infinite variety of patterns of activity that are baffling to all who attempt to catalogue or to orient them." It is impossible, therefore, to present this large group according to any one system of classification and nomenclature, and at the same time to satisfy all systematic mycologists. For the present, the best that the insect pathologist can do is to follow the lead of those who have given the greatest attention to these entomogenous fungi until some authority makes it clear that the wrong name is being used or that the fungus is in an incorrect taxonomic location. We shall attempt to follow such a procedure in the present discussion and, inasmuch as is possible, spare the reader details of the annoying controversies or the equally trying indifference that prevail among specialists in some groups of the fungi concerned. This applies to both the Ascomycetes and the Deuteromycetes (Fungi Imperfecti). In the latter group, of course, the groupings are admittedly artificial and without real phylogenetic significance.

The Ascomycetes are characterized by the formation, somewhere in the life history, of a definite number of spores (ascospores), usually eight, contained within a unicellular sac or membrane (ascus). The ascospores are formed as the result of sexual fusion of nuclei. In addition, many Ascomycetes also produce conidia, or exogenous asexual spores, borne on stalks of mycelium known as "conidiophores." As a rule, conidia are produced in large numbers when conditions are favorable for the rapid multiplication of the fungus. Many species form ascospores in only small numbers and frequently under particular and uncommon circumstances. The hyphae of Ascomycetes are generally septate, and cells thus formed are usually uninucleate.

Most authorities consider the Ascomycetes as consisting of two subclasses, Hemiascomycetes and Euascomycetes. In the first of these, which is the more primitive, the asci occur singly or in groups, but ascocarps are lacking. In the second, which includes most of the Ascomycetes, the asci are aggregated and ascocarps are present. Ascocarps are special large fruiting organs in which the asci are formed. The ascocarp may be arranged as either a widely opened, cup- or saucer-shaped structure (the apothecium), or as a hollow sphere (the perithecium) usually having a small apical opening, or ostiolum, through which the ascospores escape at maturity. Variations of these two types occur (e.g., cleistothecium and hysterothecium) but such are rare among entomogenous fungi.

**Infection Caused by Yeasts**

Of the subclass Hemiascomycetes, the order Endomycetales (Saccharomycetales), commonly called "yeasts," is known to contain a few species pathogenic for insects. As early as 1879 the Harvard entomologist Hagen advocated the use of beer mash, or what he termed a "yeast fungus," for the control of grasshoppers, potato beetles, phylloxerids, an other noxious insects. However, no actual attempt to control an insect in the field through the use of a yeast has been recorded. Several saprophytic yeasts have been isolated from insects, and a few have been described as infecting small groups of insects.

The first adequately described yeast infection of insects appears to have been that discovered in England by Keilin (1920) in larvae of the biting midge, *Dasyhelea obscuria* Winnertz, a dipteran insect living usually in the sap that fills infected wounds of elm or horse-chestnut trees. The infecting yeast, *Monosporella unicuspidata* Keil., invades the body cavity, which becomes so filled with the parasite as to give the insect's body, especially its posterior segments, a milky appearance. Even though a great number of parasites is present, the larva is able to move until is finally dies and decomposes rapidly, setting free the resistant forms of the yeast. The fat body is apparently the only organ that is completely destroyed.

The yeast itself occurs in several forms. Young organisms appear as small oval cells, from 4 to 10 microns long, budding at one end. Later, the parasites become elongated until they reach a size of about 30 microns in length by 2.5 microns in width. These elongated forms constitute the asci, in each of which develops a long needle-shaped unicellular spore with one end sharply pointed. It is assumed that this truncated end assists

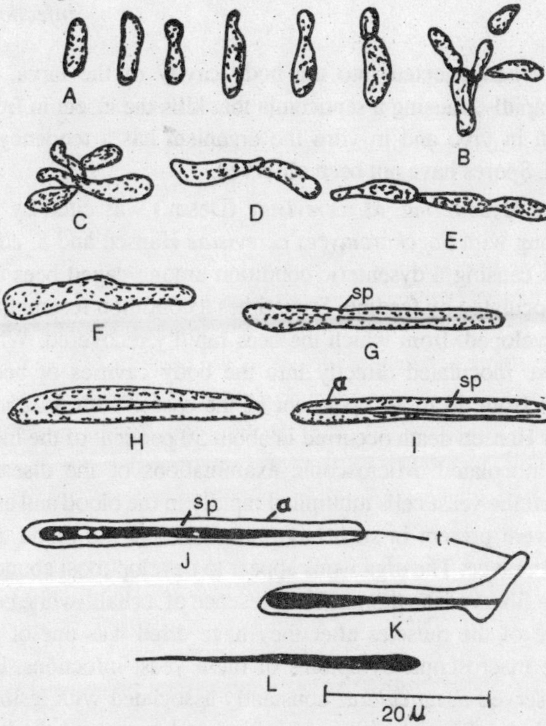


**Fig. 1.3.** *Monosporella unicuspidata* Keil. from the larva of *Dasyhelea obscura* Winnertz. *A-B* Different stages of budding cells. *B* and *C*. Rare cases of multiple budding. *D*. Ordinary budding. *E*. Chain with three cells. *F*. Elongated cell developing into ascus. *G* Ascus with beginning of spore formation. *H* and *I*. More advanced stages of spore formation. *J*. Ascus (*a*) with well-formed spore (*sp*); *t*, thickened wall of the ascus. *K*. Deformed ascus. *L*. Spore.

the spore in penetrating through the gut wall of the insect during early stages of invasion, in a manner analogous to that observed earlier by Metchnikoff in the case of a similar organism infecting a crustacean (*Daphinia*).

A yeastlike organism, which they named *Mycoderma clayi*, was isolated in 1928 by Metalnikov, Ellinger, and Chorine from European cornborer larvae shipped to them from Canada. The organism appeared as a large gram-positive rod, measuring 9 to 16 microns in length by 1.5 to 3 microns in width. Propagation occurs by budding. The yeast is readily cultivable and to a slight degree ferments glucose, levulose, sucrose, and glycerin. When injected into the body cavity of the larva, the cells multiply rapidly, causing a septicemia that kills the insect in from 2 to 5 days. Both in vivo and in vitro the organism has a tendency to form mycelium. Spores have not been observed.

Another *Mycoderma, M. cerevisiae* (Desm.) was cited by Burnside (1930) along with *Saccharomyces cerevisiae* Hansen and *S. elipsoideus* Hansen as causing a dysenteric condition among caged bees that were heavily inoculated by feeding. Sometimes a condition resembling intoxication developed, from which the bees rapidly recovered. When these yeasts were inoculated directly into the body cavities of bees, death resulted in from 50 to 100 per cent of the cases. With *Saccharomyces apiculatus* Hansen death occurred in about 50 per cent of the individuals similarly inoculated. Microscopic examinations of the diseased bees showed that the yeast cells multiplied rapidly in the blood and at the time of death were present in such large numbers that the blood took on a milky appearance. The organisms appear to develop most abundantly on the muscle fibers of the thorax. The presence of a chalk-white coating on the surface of the muscles after they have dried was one of the most distinctive macroscopic symptoms of these yeast infections. Lardinois (1926) observed *S. apiculatus* constantly associated with lesions in the tissues of bees afflicted with constipation and expressed the belief that this yeast is the sole cause of "May disease."

In an attempt to find a microorganism that might be used in the control of mealybugs in Russia, Evlakhova (1939) reared *Pseudococcus citri* Risso in the laboratory under conditions of high humidity to favor the development of disease organisms. Among those isolated was a yeastlike fungus which he named *Blastodendrion pseudococci*. Under experimental conditions of 25°C. and 60 to 70 per cent humidity, a mortality of mealybugs occurred ranging up to 50 per cent; with cultures of increased virulence a mortality of as high as 100 per cent was reached. Most of the mealybugs died in 24 hours after commencing to feed on potato stems smeared with the cultures. Spraying the mealybugs with a suspension of the organism was less effective. In causing the infection, the yeast apparently entered the insect's body through the digestive tract, penetrating into the fat body and the muscular system.

An interesting pathology involving a yeast has been described by Frobisher (1926) in *Drosophila melanogaster* Meig. being reared in the daboratory. The yeast concerned was a red torula, which was being used as food by the drosophila flies. Occasionally there appeared in the flasks used for cultivation a blue-green *Penicillium* which was capable of infecting and killing the flies. Besides the actual invasion of the insect tissues by the fungus, another mechanism seemed to be responsible for the death of the insects. The fungus appeared to grow among the torula cells in the gut and by matting them together formed a more or less solid mass that could block both elimination and engorgement. By invasion of the intestinal wall, the fungus hyphae were believed to bind this mass firmly in the intestine.

## HYPOCREALES INFECTIONS

The order Hypocreales belongs to the subclass Euascomycetes and contains several genera noted for their entomogenous species. These genera fall into rather heterogeneous groups, which are nevertheless convenient for purposes of discussion. It is for this reason that we shall consider as a group the fungi parasitic on scale insects and those parasitic on whiteflies, even though some of the fungi coming under these categories are not Hypocreales. Some Hypocreales, such as *Cordyceps*, are clearly defined as taxonomic groups and may be discussed as such.

### Cordyceps Infections

The genus *Cordyceps* contains about 200 known species, nearly all of which parasitize insects. Almost 40 recognized species occur in the United States. These have been studied principally from a taxonomic standpoint. Some species originally considered as *Cordyceps* are placed by some authors; (but not now recognized by Mains, 1948) in the genus *Ophiocordyceps* which differs from the former in having clavate asci and fused overlapping ascospores. Cordyceps are cosmopolitan in distribution and occur on representatives of several orders of insects, principally Hemiptera, Diptera, Lepidoptera, Hymenoptera, and Coleoptera. The larval stage appears to be the most frequent host, but different species may occur on every stage of insect development. Information concerning the identity of the various hosts is frequently meager because of the destructive or obliterating effect the fungus has on the insect and because of the tendency early mycologists had of reporting the fungus simply on a "caterpillar," "pupa," "moth," and the like.

Associated with the genus *Cordyceps* are numerous interesting historical records. Early writers have presented rather extensive accounts of the group as it was known to them at that time. Insect parasitized by these fungi were frequently known as "vegetable wasps" and "plant worms" (or "awetos" in New Zealand). Among the most celebrated of the "vegetable wasps" were those (*Vespa* and *Polyna*) parasitized by *Cordyceps sphecocephala* (Klotzsch). An account exists in which Torrubia, a Franciscan friar, tells of finding, in 1749, some dead wasps in a field near Havana and "from the belly of every wasp a plant germinated, which grows about five spans high." As recounted by Gray, Torrubia gave a representation of two wasps lying on the ground with a tree growing out of the base of each abdomen, while three other wasps are flying (!) around these trees, each flying insect having a similar tree affixed to it. These "trees" in all probability represent species of *Coryceps*. On this basis it may be considered that the earliest known record of an entomogenous fungus was that made by Christian Paulnus in the beginning of the eighteenth century when he wrote that "certain trees in the island Sombrero in the East Indies have large worms attached to them under ground, in the place of roots." For additional curios of this type of historical information the reader is referred to Gray's (1858) privatey printed "Notices of insects That Are Known to Form the Bases of Fingoid Parasites."

*Morphological Characteristics*. The feature that characterizes the genus *Cordyceps* is the fact that the stroma arises from a sclerotium formed within the body of the insect on which the fungus is parasitic. As described by Massee (1895), who monographed the genus, this stroma usually consists of an erect stemlike sterile portion composed of a fascicle of irregularly parallel hyphae, white internally, the external or cortical hyphae being usually tinged the color and in many species giving off numerous short later branches that form the minutely velvety or downy exterior of the stem.

The "head" or "club", which is the fertile portion of the stem, is usually terminal in location and may be brightly colored. The perithecia originate side by side deep in the stroma, their shape ovate or flask-shaped, their mouths reaching the surface of the stroma. They may remain completely immersed or at maturity be quite superficial, with the entire perithecium being exposed. The latter condition causes the surface of the head to be rough, whereas when the perithecia are immersed it is smooth.

The asci contain eight spores and are very long and slender. As the spores mature, the contents of the head become swollen, and the wall of the ascus is ruptured at the apex. The slender spores, arranged in a parallel fascicle, are almost as long as the ascus, are hyaline and multiseptate or continuous. After escaping from the ascus, the multiseptate spores usually break up into their component cells, which eventually germinate. Paraphyses are absent.

Some species of *Cordyceps* are believed to have their conidial stages in such genera as *Isaria* and *Botrytis*. Such supposed relationships have not, however, been made too clear. For example, *Isaria farinosa* (Dicks) (or *Spicaria farinosa* (Fron.)) was for years considered to be the conidial stage of *Cordyceps militaris* (Lk.). This belief and its supposed proof were discounted by Petch (1936), who contends that the conidial stage of this *Cordyceps* is a *Cephalosporium*. *Isaria farinosa* is a fasciculate *Spicaria* and may occur as an *Isaria* or as a simple *Spicaria*. It has no relation to *Cordyceps militaris*, which attacks larvae and pupae of Lepidoptera (and possibly Coleoptera;), while *Isaria farinosa* is a general entomophyte occurring on Lepidoptera, Hymenoptera, Coleoptera, Diptera, Aphididae, and Arachnida. As Petch pointed out earlier (1934), the genus *Isaria* is an unsatisfactory one from several standpoints. If it is retained, it is a compound *Spicaria*. Petch suggests that it would be simpler to discard the genus *Isaria* and retain the name as purely a descriptive term. A relationship between the imperfect genus *Hirsutella*, once considered a basidiomycete, and certain species of *Cordyceps* has been suggested by Speare (1920), and supported by Petch (1923), who reduced his genus *Trichosterigma* to synonymy with *Hirsutella*. Since then several species of *Hirsutella* have been found to be the conidial stages of species of *Cordyceps*. Such may eventually be found to be the case with other species of *Hirsutella*, but many identities as to their ascigerous stages still remain uncertain (e.g. *H. subulata* Petch on the larva of the codling moth, *Carpocapsa pomonella* Linn., and *H. saussurei* (Cke.), a parasite on hornets in the tropics and others). Other genera of Hypocreales may, however, be related to *Hirsutella*. Species of *Calonectria*, for example, parasitize leafhoppers and are the ascigerous stage of certain species of *Hirsutella*.

The first to describe the growth of any species of *Cordyceps* from spore to spore in the laboratory appears to have been Shanor (1936), who accomplished this with *Cordyceps militaris*. Living lepidopterous pupae were inoculated with hyphae and placed in moist sterile sphagnum moss. Normal mature fruiting bodies were consistently produced. No success in producing perithecia has been obtained using artifical culture media.

*Pathogenesis.* To illustrate the life history of most *Cordyceps* and the manner in which they infect their hosts, we might use, as did De Bary (1887), the well-known *C. militaris* as an example.

After being ejected from a ascus of the orange-colored club-shaped stroma, the slender filiform, or rod-shaped spore, divides transversely into a large number of secondary spores. When one of these lands on the slightly moist skin of the host caterpillar, it swells slightly, becomes rounded in shape and puts out a germ tube. Sometimes the spores become partly united again by means of short connecting tubes before they germinate.

The germ tube proceeds at once to penetrate the integument, within which it enlarges into a somewhat thicker fungous hypha which ramifies its way into the deeper layers of the skin. It finally breaks through into the body cavity of the insect, insinuating itself between the muscular and fat tissue. Here the hypha breaks down into cylindrical bodies ("cylindergonidia") similar to the hyphal bodies already described in the case of the entomophthoraceous fungi. The continue to pass into the blood of the insect, where the elongate to twice or several times their original size, dividing repeatedly by transverse walls and by terminal and lateral sprouts. These cells disperse through the blood of the host until the body cavity gradually becomes filled with them as the quantity of blood is diminished. At this point the larva loses its normal turgidity, becomes soft and relaxed, and dies.

As soon as the insect dies, the sprout cells begin to develop rapidly at the expense of the dead body tissues, forming branched hyphae, which fill the entire body cavity except the alimentary canal and expand the body to its former size and turgidity. The various tissues of the body are more or less absorbed by the fungus, and in 1 or 2 days a body is formed that retains the approximate size and shape of the living insect but consists essentially of a mass of fungous hyphae with some remains of the insect. This fungus body has all the properties of a sclerotium and may be considered as such. If this sclerotium lies in a moist situation, it can give rise directly to fresh stromata; if dried, it may lie dormant for several months.

The development of most *Cordyceps*, in their simplest form, follows much the same course as the one just described by *C. militaris*. Nevertheless variations occur between species and, in fact, within the same species. The life histories of species of *Ophiocordyceps* in most respects are similar to those of *Cordyceps* species.

**Fungi Parasitic on Scale Insects**

Some of the most notable work on the fungous diseases of insects has concerned those fungi parasitic on scale insects. Not all these fungi belong to the order Hypocreales, although many of the most important ones do. Some are known only in their imperfect or conidial stages. Since those which do belong to Hypocreales have been favored with a great amount of detailed study, it is considered convenient to use this order as a center about which all the fungi parasitic on scale insects may be considered.

According to Petch (1921), the earliest record of a fungus parasitic on a scale insect was made in 1848 by Desmazières, who collected his specimens from willow and ash at Caen, France. For many years following this, mycologists (including such systematists as Berkeley, the Tulasnes, Saccardo, and Petch) concerned themselves almost exclusively with the taxonomic aspects of these fungi. Eventually it was realized that at least some of these fungi may constitute an important natural check on certain species of scale insects. Beginning about 1912 considerable attention was focused on them in the citrus-growing areas of Florida, and attempts were made to use the fungi in the biological control of the insects. For a time it was believed that the increase of scale insects following Bordeaux spraying was the result of killing off the entomogenous fungi. That this was probably not entirely the case is indicated by the observations of a number of men, including Holloway and Young (1943), who attributed the increase to the effect of the spray residues. Apparently a complex of several fluctuating factors is involved. There does appear to be a need, however, to redetermine the true pathogenicity of at least certain of the entomogenous fungi found on scale insects. That some of these fungi are actually saprophytes or secondary parasites is a distinct possibility. In fact, Fisher (1947) in her studies of certain of the entomogenous fungi (*Microcera*, *Podonectria*, and others) of scale insects in Florida makes the statement that "No evidence has been found that any of the so-called 'friendly fungi' actually parasitize any of the scale insects."

The fungi found on scale insects belong to a number of genera; and since many of them are important enough to deserve more than mere mention, it seems advisable to consider them according to their generic groupings, in much the same manner as that employed by Fawcett (1948). Some of the species discussed have at one time gone under other names; some have several synonyms which, for reasons of space, cannot be listed here.

*Sphaerostilbe Infections*. Three important species of *Sphacerostilbe* are found associated with scale insects: *S. aurantiicola* (Berk. & Br.), *S. Flammea* Tul. and *S. coccidophthora* (Zimm). The first two species are found on insects in North and South America, the Orient, and in other parts of the world. *S. coccidophthora* occurs in the Orient on coffeetea-, and bamboo-scale insects. They are commonly called the "redheaded scale-fungi."

*S. aurantiicola*, which Snyder and Hansen (1945) have designated as *Nectria episphaeria* f. *coccophila* (Desm.) (= *S. coccophila* Tul.; *N. coccophila* (Tul.), occurs on scale insects which live on citrus and other host plants. In North America, principally in Florida and the West Indies, the best known hosts on citrus include California red scale, Florida red scale, Spanish red scale, Putnam scale, ivy scale, snow scale, green scale, purple scale, thread scale, Glover scale, chaff scale, black scale, rufous scale, and San Jose scale. It has also been reported on the citrus mealy-bug. It appears to be one of the most prevalent fungi on scale insects, especially in Florida, where it is reported to have more "effect" on the purple than on the Florida red scale—the opposite of the pink fungus, *Nectria diploa* Berk. & Curt., which appears to be more, "effective" against the red scale. The perithecia are small, globose, and orange red to blooa red in color. The imperfect pustules are of the same color but are clavate or flattened pulvinate in form. The fungus has been obtained in culture from both conidia and ascospores. Working with the obscure scale, *Chrysomphalus obscurus* (Comst.), Luttrell (1944) observed that the fungus completely destroys the body of the infected scale and forms a plectenchymatous stroma between the shield of the scale and the bark of the plant host. The bark is not penetrated. On the other hand, Fisher (1948) reports that all her attempts to artificially inoculate red and purple scales with the fungus have failed.

*S. flammea* has been reported on the snow scale on citrus and on scale insects of many plant hosts besides citrus. The heads of the imperfect form of this fungus are red in color; those of the perfect form are bright orange-red. The latter are crowded together on a stroma which often hides the scale insect.

The imperfect, or conidial, stages of *Sphaerostilbe* are generally placed under the generic designation *Microcera*, which, according to some authorities, is synonymous with *Fusarium*. Thus Snyder and Hansen (1945) have provided the combination *Fusarium episphaeria* f.

*coccophila* for the imperfect stage of *Nectria episphaeria* f. *coccophila* (*S. aurantiicola*). Petch, however, still recognizes and uses *Microcera*.

*Nectria Infections*. In the United States two species of *Nectria* are of importance: *N. diploa* Berk. & Curt., and *N. vilis* (Syd.). Concerning the latter, only the imperfect form *Tubercularia coccicola* Stev. has been reported from North America, where it occurs on the snow scale, purple scale, and probably others. *N. barbata* Petch occurs on scale insects on citrus in Ceylon.

*Nectria diploa*, commonly known as the "pink fungus" in Florida, where it was discovered in 1912 forms pustules somewhat similar to those of *Sphaerostilbe*. The stroma is light rose or pink in color. A pink or reddish border surrounds the scale on which it has grown. It occurs on numerous scale insects in various parts of the world. It has its imperfect stage in *Microcerca* (or Petch's *Pseudomicrocera*). The Florida red scale has been reported to be particularly susceptible to its action, and Watson (1913) claimed success in spreading it artificially to good effect. The fungus also occurs on purple scale but with less frequency than on the red scale.

*Podonectria, Lisea, and Torrubiella Infections*. In 1921 Petch proposed the generic name *Podonectria* (= *Berkelella*) to include three species of fungi commonly called the "white-headed scale-fungi": *P. coccicola* (E. & E.), *P. aurantii* (Hoeh.), and *P. echinata* Petch. All are found on scale insects.

*P. coccicola* was one of the first entomogenous fungi observed in Florida, where it occurs on Glover scale, purple scale, and chaff scale. It has also been found in South America, the Orient, and elsewhere. Watson and Berger (1937) are of the opinion that it was this fungus which saved the citrus industry of Florida in the 1830's when Glover scale (or the "long scale") was introduced into that state. The trees were at first killed back each year by the scale until presumably the fungus came to the rescue and saved them. Other feel that the fungus is in reality only a secondary parasite or a saprophyte which grows on the armors of the scales. Fisher (1947) observed in growing in such a location on purple scale that had been killed by the endoparasitic chytrid, *Myiophagus* sp. It usually appears as small whitish heads growing out from the scales. The imperfect form is *Tetracrium coccicolum* Hoeh.

*P. aurantii* is similar in appearance to *P. coccicola* and is found in Brazil on citrus and in Formosa on the black parlatoria scale. *P. echinata* occurs in Ceylon on *Lepidosaphes* sp.

*Lisea parlatoria* Zimm. has been observed on the black parlatoria scale on citrus leaves in Java. The perithecial heads are dark violet to black in color.

*Torrubiella lecanii* Johnst. occurs on the hemispherical scale in Cuba and Puerto Rico. Its perithecia are of a vivid yellow color. Incidentally, at least three species of *Torrubiella* have been reported from North America. One species, *T. gibelludae* Petch, a parasite of arachnids, is known to be the perfect stage of *Gibellula aranearum* (Schw.). A considerable number of species originally placed in the genus *Torrubiella* are in reality members of other genera or synonyms of other species.

*Hypocrella and Aschersonia Infections*. Member of the genus *Hypocrella*, included in the genus *Hypocrea* by early mycologists, have their imperfect stages in *Aschersonia.* The genus *Aschersonia*, as such, will be treated more fully in our section dealing with the fungi parasitic on whiteflies.

As has been pointed out by Petch (1921), even though species of *Hypocrella* so closely resemble the corresponding species of *Aschersonia* that it is not possible to decide which a given stroma is without sectioning it, yet it was apparently not until 1896 that any relation between the two was suggested. Since then it has definitely been ascertained that *Aschersonia* is the conidial form of *Hypocrella*. Nevertheless, the *Aschersonia* stage is found much more frequently than the *Hypocrella* stage. An *Aschersonia* stroma usually does not subsequently become perthecial although there are exceptions to this. In some gatherings all the stromata will be *Hypocrella*; in others all *Aschersonia*. Just what conditions govern the production of either stage is not known. Occasionally both stages may be found in the same stroma. It is by such findings that it is possible to correlate species of *Aschersonia* with their *Hypocrella* stages.

In 1921 Petch wrote that 70 species of *Hypocrella* and 60 species of *Aschersonia* have been described. Of these, however, only a total number of 54 names covering 42 species are valid. In the group found on scale insects there were then known to be 20 species of *Hypocrella*; the corresponding *Aschersonia* was known in 11 cases. In 1947 Petch informed the writer that because certain material was not available to him during World War I, these figures should actually have been slightly different. At any rate, in the scale-insect group of fungi, Petch states that there are now known to be 22 species of *Hypocrella*, the corresponding *Aschersonia* being known in 12 cases, and 3 unattached species of

*Aschersonia.* Charles (1941) lists 8 species of *Hypocrella* known in the United States, at least 5 of which have known *Aschersonia* stages.

Whereas species of *Sphaerostilbe* and *Nectria* occur on the armored scales (Diaspididae), species of *Aschersonia,* on the other hand, infect soft scales (Coccidae) and whiteflies (Aleyrodidae). In the latter case, there is also a difference between the two families in that those *Aschersonia* parasitic on whiteflies have paraphyses (elongated sterile cells) in the pycnidium, while those parasitic on the coccids have no paraphyses. It might further be mentioned that considerable care should be exercised when examining scale-infested trees to decide just which insect a *hypocrella* is infecting. If an armored scale and a soft scale occur together on the same plant, the fungus may destroy all the soft scales leaving only the armored scale. Cursory examination may then lead the observer to think that the armored scales he sees were the hosts of the *Hypocrella.*

As far as scale insects are concerned, the better known *Hypocrella* and their *Aschersonia* counterparts include *Hypocrella epiphylla* (Mass). (*Aschersonia cubensis* Berk. & Curt.), *H. turbinata* Berk. (*A. turbinata* Berk.), and *H. javanica* (P. & S.) (*A. coffeae* P. Henn.).

*Myriangium Infections.* Although not a Hypocreales, but of the order Myriangiales, the genus *Myriangium* merits consideration at this point because it is one of the important parasites of scale insects throughout the tropics. Modern descriptions of the genus have been presented by Petch (1946) and by Miller (1940). In the words of the latter author, the plant body of *Myriangium* consists of a black pseudoparenchymatous stroma, which later gives rise to a peculiar type of apothecium with an apically delimited fertile region, consisting of asci at different levels embedded in coalesced fungous tissue. Apparently there is no conidial stage. The stroma is superficial on the bark of the tree, and under each stroma are several dead scales, penetrated and covered by mycelium. There is a definite relationship between the death of the scale and that of the limb and the *Myriangium.* Live stromata are not found on dead branches.

The commonly known species of *Myriangium* include *M. duriaei* Mont. & Berk. (*M. curtissi*) (Mont. & Berk), *floridanum* Hoch., and *M. montagnci* Berk. Several authors have cited these species to be effective natural control agents, especially against citrus scales.

*Other Fungus Parasites of Scale Insects.* In addition to the species of Hypocreales and Myriangiales mentioned in the preceding paragraphs, a number of fungi parasitic on scale insects exist in other groups of fungi. Most of them remain among the Fungi Imperfecti.

*Cephalosporium lecanii* Zimm. occurs on several species of scale insects in the Americas and in certain islands of the South Pacific. In Florida it has been found on a number of scales on citrus and from a natural-control standpoint has been reported to be an effective parasite of the pyriform scale. In Brazil it is considered to be important in the control of the green scale on coffee plants, where it affects all the internal parts of the insect, including the eggs. The fungus grows around and over the scale insects. It is at first white but finally becomes pale yellow or lemon yellow in color. It usually has a powdery or mealy appearance, which is due to the numerous minute heads or spore clusters that develop on the conidiophores breaking out through the scale. Eventually it may appear smooth and waxy because of the fusion of the conidial heads. The fungus grows well on artificial media. Some authorities consider this fungus to belong in the genus *Verticillium* along with another species, *V. cinnamomeum* Petch, which is found on scale insects and whiteflies in Florida. Other species of *Cephalosporium* have been reported on scales as well as other insects in countries other than the United States.

Similar to *Verticillium* is the genus *Cladobotyrum*, which contains at least one species, *C. heterocladum* (Penz.), parasitic on scale insects. It was described originally as *Verticillium* on the brown scale in Italy, and this or a closely related species has been so designated in Florida.

Several species of *Fusarium* have been reported on scale insects in various parts of the world. *Beauveria globulifera* (Speg.), the so-called "chinch-bug fungus", is sometimes found on the brown scale. *spicaria javanica* Bally occurs on several scales in Ceylon as well as on the cottony cushion scale there and in Florida, where its control value has been noted when conditions were optimum. *Aspergillus depauperatus* Petch (considered to be a strain of the *A. restrictus* Smith series not far from *A. gracilis* Bainier) has been identified on specimens of California red scale on citrus in Palestine, and on other species in England and Ceylon. In certain parts of the world representatives of the genera *Pegiotrichum, Acrostalagmus*, and *Rhinotrichum* are found parasitizing scale insects. Species of *Cladosporium* are also found in this relationship but probably only as saprophytes or weak parasites.

In moist seasons along the California coast, a white powdery fungus appears on the black scale, sometimes killing considerable numbers of them. This fungus has been known as "*Isaria*" although no true *Isaria* stage has been observed. Quayle and Taylor (1915) were able to kill a fair percentage of scales with this fungus in moist chambers in the laboratory, but attempts to initiate outbreaks in the field were unsuccessful. Incidentally, as early as 1898 attempts were made to control the black scale in California by means of fungi, with favorable results being reported.

Although on the whole the relationship between scale insects and species of *Septobasidium* (Basidiomycetes) is one of mutualistic symbiosis, enough of any colony of scale insects are parasitized to warrant the mention of the genus along with other fungi parasitic on this group of insects.

**Fungi Parasitic on Whiteflies**

A considerable number of species of fungi are parasitic on the immature stages of several species of whiteflies (Aleyrodidae). Particular attention has been given to these combinations of parasitism in Florida, where the two most prominent species of whiteflies on citrus are the citrus whitefly, *Dialeurodes citri* (R. & H.), and the cloudy-winged whitefly, *Dialeurodes citrifolii* (Morg.). At least five species of fungi occur commonly on the nymphs of these two species. These are of enough importance to merit separate discussions.

*Aschersonia aleyrodis Webber*. The so-called "red aschersonia", *Aschersonia aleyrodis*, occurs on a number of species of whiteflies other than its principal hosts in this country. The fungus has its perfect stage in *Hypocrella libera* Syd. The imperfect stage forms a raised, flattened, pulvinate, red to pinkish-buff, stromatic pustule. It was first described in 1897 by Webber, who observed that, although all immature stages of the insect are readily attacked, those attacked while young were most abundant. In the area of Panasofkee, Florida, he noticed that the whitefly population decreased markedly with the spread of the fungus. In previous years the whiteflies and the accompanying sooty mold had been so abundant as to require the washing of all fruits. A few years after the fungus had extended its effectiveness no fruits had to be washed.

Soon after a nymph becomes infected with the fungus, it becomes swollen and may secrete more honeydew than usual. Within the aid of a lens the intricately woven hyphae may be seen within the body of the insect. As the fungus develops internally, the interior organs appear to contract away from the margin. Soon after this the insect dies and the hyphae break through the body wall and form a dense marginal fringe around the edge of the insect. Subsequently the red pustual with its spores develops. As Faweett (1948) has pointed out, what is often referred to by entomologists as "natural" or "unexplained mortality" may in reality be caused by the fungus in the early stages of its growth within the body of the insect. If conditions are such that the nymphs should become completely dried soon after death, the development of the fungus would be stopped before the latter became visible and in this way pass for "unexplained mortality."

The fungus was first cultivated by Faweett (1908), using sweet-potato strips. Berger (1921) used this method in modified form for the production of large quantities of the fungus by inoculating sterilized sweet-potato slices in pint bottled. Matured cultures were obtained in this way in from 30 to 40 days and could be held in storage for extended periods of time. These bottles were than sent out to the growers who, after shaking the culture in water and filtering the suspension through cheesecloth, distributed the fungus artificially by spraying. One pint of culture was recommended for each acre of the orchard. The season in which the most successful results could be expected was during the rainy moist-weather period of June and July—Florida's rainy season.

The grower did not necessarily have to depend upon these cultures for a supply of the fungus. This could usually be secured by obtaining infested leaves from another orchard. Such leaves, bearing the fungus-infected insects, could be placed in water and the resulting spore suspension used as a spray. Before the spray method was well understood, leaves bearing pustules of the fungus were pinned to trees infested with whiteflies. This method, though reasonably effective, was so slow that the spray method soon replaced it.

*Aschersonia goldiana Sacs. & Ellis*. The yellow aschersonia, *A. goldiana*, for a number of years was identified in Florida as *A. flavocitrina* P. Henn. It has been reported on species of whiteflies in Florida, the West Indies, Panama, and on an unidentified insect in Venezuela. In Florida it is parasitic especially on the cloudy-winged whitefly, *Dialeurodes citrifolii* (Morg.), although it is occasionally found on other species.

*A. goldiana* resembles the red aschersonia in habits and general appearance, except that its pustules are yellowish in color. It can be grown on sweet-potato slices in much the same way as that used for the red aschersonia. It does not appear to be so effective a control agent as was first expected and is useful only in combating the cloudy-winged whitefly, for which it is probably no more effective than the red aschersonia. For interesting field reports on this and other entomogenous fungi, the reader is advised to consult the Annual Reports, as well as other publications of the Florida Agricultural Experiment Station during the early 1900's.

*Aegerita webberi Faw.* This fungus, commonly called "Webber's brown fungus", occurs in the United States, the West Indies, Ceylon, India, and New Zealand. In Florida its principal hosts are the *Dialeurodes* species mentioned earlier and *Aleurothrisus howardi* (Quaint). In Cuba the citrus blackfly, *Aleurocanthus woglumi* Ashby, is attacked. The fungus was first discovered in its sterile form in 1896 by Webber; its spore stage was found by Faweett (1910), who named it *Aegerita webberi*. It is a deuteromycete, the perfect form not having been found.

In its sterile form the mature fungus consists of a chocolate-brown compact stroma covering the entire insect, from the margin of which extend colorless thick-walled hyphae. Later in the development of the fungus, usually during the summer or fall, the marginal hyphae grow out long and colorless, extending not only over the undersurface of the leaf but also around the edges and upon its upper surface. The hyphae may sometimes extend down the petiole and along the stem to the next leaf. Every larva in the area covered becomes infected. When the hyphae become very abundant, the fungus may almost entirely cover the leaf surface with silky grayish-brown strands. The sporodochia, which consist of aggregations of conidialike, inflated, spherical cells, are borne on the upper surface of the leaf, on short lateral hyphae. These aggregations usually remain in union and function as a spore. Their light weight and radiating appendages aid considerably in their dissemination by the wind.

The fact that *A. webberi* may spread by means of superficial hyphae spreading over the surface of the leaves, as well as by sporelike aggregations of cells carried by air currents or by insects, reportedly makes it one of the most efficient parasites of whitefly nymphs, provided that conditions for growth are favorable. Its ability to spread rapidly was one of the characteristics noticed by Webber when he first found this fungus in part of a 5-acre orchard in Manatee, Florida. At that time millions of live nymphs could be found in the trees of this orchard where the fungus had not spread. There was no trace of the fungus in adjoining orchards. Nine months later the fungus had spread so rapidly that it was difficult to find a living specimen of the whitefly in this 5-acre orchard. Trees in the adjoining orchards remained heavily infested with the insect. Within an additional year's time, the fungus had spread over a radius of approximately 2 miles, killing all immature stages of the insect as it spread. Although the accuracy of these observations had been questioned, Webber himself did not see fit to alter his opinion. That the sporodochia themselves might spread the fungus was shown experimentally by Fawcett (1910) when he inoculated nymphs by drawing over them a camel's-hair brush moistened with water containing the aggregated sporelike cells. The insects showed signs of infection in 9 days, and in 16 days the stromata burst through the edges of the nymphs.

Since no satisfactory method of growing *A. webberi* in pure culture in large quantities has been developed, its distribution is usually effected by one of the three following methods: (1) pinning onto leaves containing whitefly larvae other leaves bearing infected insects and mature pustules; (2) washing off or triturating the pustules in water and spraying this suspension on the infested trees; (3) planting young fungus-bearing trees in such a way as to have their leaves intermingle with the whitefly-infested trees. In certain areas and under optimum conditions satisfactory control of whitefly larvae was obtained using these methods. Natural outbreaks, especially in certain low-laying hammock groves, are sometimes equally effective, making artificial remedial measures unnecessary.

*Fusarium aleyrodis Petch.* This fungus has already been referred to in connection with scale insects. It is best known, however, for its parasitization of whiteflies, *Dialeurodes citri* (R. & H.), and *D. citrifolii* (Morg.). Because it presents a delicate fringe of white hyphae growing outward from the edges of the larvae, *Fusarium aleyrodis* is commonly known as the "white-fringe fungus." It is probably the best known of about six identified and a considerable number of unidentified entomogenous species of *Fusarium* in North America.

The hyphae at first bear one-, two-, or three-celled conidia, oval to fusiform in shape. Later pinkish spore masses are formed on the edge of the larvae. The fungus is readily cultivable in pure culture.

The role of this fungus in the control of whiteflies in Florida does not appear to be so definite as is the case with some of the others. Under certain conditions, Watson (1913) found that 98 per cent of the larvae showing a "natural mortality" were infected with the fungus even though it was not apparent to the unaided eye. Adult whiteflies were also observed to harbor the fungus, which Watson succeeded in isolating from them. The eggs of the insects apparently also become parasitized and destroyed by the fungus.

*Other Fungi Parasitic on Whiteflies.* Similar in general appearance to the pustules of Webber's brown fungus are those of the cinnamon fungus, *Verticillium cinnamomeum* Petch, first seen in Florida in 1905. It has been found on both whiteflies and scale insects. Several species of *Aschersonia* and *Hypocrella* have been reported on aleyrodids in various parts of the world, including the West Indies and the Orient. According to Petch (1947), in the whitefly group of fungi, there are 13 species of *Hypocrella* of which the Aschersonia stages are known, and 14 unattached species of *Aschersonia*. Another Hypocreales, *Stereocrea aurantiaca* Petch, occurs on whiteflies in Florida and is the perfect stage of an *Aschersonia*. In Ceylon, *Rhinotrichum album* Petch has been observed on a species of aleyrodid.

An unidentified species of *Sporotrichum* has been isolated from the adult winged stage as well as from the nymphs of whiteflies in Florida, and it was thought to be responsible for the death of large numbers of them.

**Other Ascomycete Infections**

In the past some authors have considered many of the entomogenous fungi as belonging to the order Sphaeriales. Modern treatments, however, clearly distinguish between the two orders. Both are extremely large groups, but as now constituted, most of the entomogenous species fall in the order Hypocreales. Only a few are included in the Sphaeriales. The entomogenous members of some genera, e.g., *Sphaeria*, have been placed in synonymy with certain Hypocreales, e.g., *Ophicordyceps*.

As far as North America is concerned, the remaining orders of Ascomycetes contain relatively few species of fungi definitely pathogenic for insects. Species such as *Scorias spongiosa* (Schw.) in the order Dothideales have been found associated with insects but are probably saprophytic in nature, since the initial substratum frequently consists of insect secretions. A species of *Eurotium* (order Eurotiales) has been listed by Charles (1941) on a chalcid. Some primitive Eurctiales grow saprophytically on the pupal cases of certain Lepidoptera. Since this order includes many species of *Aspergillus* and *Penicillium*, ordinarily known in their imperfect or conidial stages, it is entirely possible that some of these entomogenous Fungi Imperfecti will eventually be grouped here. The genera *Cenangium* and *Scelerodenris* of the order Helotiales have at times been used to include entomogenous members, most of which have now been placed in synonymy with other Ascomycetes.

Although members of the large and important order Laboulbeniales may be considered as cutaneous parasites, they are generally thought of as commensals.

**The Muscardine Diseases**

Since the term "muscardine" has been indiscriminately used in entomological literature to mean almost any type of fungous infection, it appears desirable to consider the origin and proper use of this word.

As far as its modern usage is concerned, the term "muscardine" apparently originated in the Italian language with the word "*moscardino*", meaning a musk comfit, grape, pear, and the like, or any of the various plants with musk-scented foliage or flowers. ("Musk", incidentally, refers to the odorous substance from the abdomen of the male musk deer, used as a basis for perfumes.) The French have the words "*muscadin*", meaning a musk lozenge, and "*muscardin*", which, in addition to refererring to the dormouse (*Muscardinus*), also means a comfit or bonbon. Because the bodies of the insects infected with the fungi we are about to describe are transformed into white mummified specimens resembling in appearance comfits or bonbons, the natives of France referred to them as "*musardin.*" French scientists added a final "e" to the word and used it in referring to the fungus concerned. It has since been taken over as a bonafide English word, and English dictionaries and encyclopedias furnish us with at least three meanings or uses of the word: (1) as a noun meaning the fungus; (2) as a noun meaning the disease caused by the fungus; (3) as an adjective, "muscardined."

The word "muscardine" was first used as it applied to the well-known disease of the silkworm, and also referred specifically to the fungus *Beauveria bassiana* (Bals.). Soon thereafter it was also used in reference to the fungus we now know as *Metarrhizium anisopliae* (Metch.), and because of the green color of the spores it was called "green muscardine." In addition to these two infections, it might be permissible to use the term generally in connection with those mycoses in which the fruiting bodies arise on the exterior of the insect, producing a thick covering about the animal. The fungi concerned are known primarily in their imperfect stages (Fungi Imperfecti).

**Muscardine of the Silkworm**

Muscardine of the silkworm, *Bombyx mori* (Linn.), occurs throughout the world wherever this insect is reared. The disease has been particularly important in France and Italy, where tremendous losses have been experienced. Even after the nature of the muscardine and its control were understood, losses were considerable. For example, in northern Italy, around 1925, approximately 11 million pounds of cocoons were being lost every year from muscardine alone. Outbreaks still occur, but for the most part they are sporadic. In Italy, incidentally, the disease was first called "*mal del segno*" (the disease having a sign); later and at present it is known as "*calcino*" (calcium; white powder).

*Early History*. It is not strange that muscardine was the first acknowledged disease of the silkworm, since it is so easily recognized. The white mummies into which the diseased larvae are transformed are very distinctive and noticeable, and early sericulturists were able to ascertain their presence readily.

Until 1835 it was generally believed that the disease was not of a contagious nature but that it was caused by a variety of agencies, notably those pertaining to meteorological conditions and rearing techniques. For example, Boissier des Sauvages (1763) believed that it was due to a particular state of the atmosphere which precedes storms and which is called "*touffe*" (i.e., "wisps of heat"). Nysten (1808) attributed the disease to defective incubation. Dandolo (1825) declared that it resulted from abnormal physiological conditions. The white efforescence that develops on cadavers was, according to him, "the original mineral." Other causes were postulated by other writers, but most of them were variations of these amicrobic influences.

The credit for first showing that the disease is a contagious one and that it is parasitic in origin is generally given to Bassi de Lodi (1835–1839), who showed that the affliction was caused by a fungus that multiplied in and on the body of the silkworm. This discovery naturally created quite a sensation among sericulturists and microbiologists alike. Other investigators were drawn to this interesting case of parasitism. The fungus itself was studied and described by Balsamo, who gave it the name *Botrytis bassiana*, the specific name honoring Bassi. Audoin (1837) elaborated on Bassi's discovery and demonstrated that this "cryptogam" would reproduce the disease when inoculated artificially into the body cavity or the fatty tissue of a healthy silkworm. This was true regardless of the age or instar of the caterpillar. In order to ensure the completion of the fungus's development, however, certain humidity requirements were necessary. Audoin also observed that the malady was not peculiar to the silkworm but that many other insect species were also susceptible. In 1839 Johanys was able to cultivate the fungus on nonliving organic media.

The contagiousness of muscardine was not, however, accepted by all biologists and sericulturists without question. It took some years before the relation between the fungus and its host was thoroughly understood. Such a recognized authority as Guérin-Méneville (1848), for example, doubted the contagiousness of the infection and believed that the manifestations of the fungus were merely symptoms or signs of careless handling. Vittadini (1853), however, did a satisfactory job of refuting Guérin-Méneville's arguments and of showing that the fungus was, in fact, the true cause of muscardine.

Among the more modern workers (i.e., since 1900) on this silkworm disease or on the fungus itself are included such names as Quajat, Verson, Conte and Levrat, Beauverie, Arnaud, Paillor, and Masersa. These and others have contributed to our knowledge of muscardine until now it is a well-understood infection and one that can be coped with intelligently.

*The Causative Fungus.* As we have already mentioned, Balsamo placed the fungus responsible for muscardine in the genus *botrytis* of which *Botrytis cinerea* Pers., the cause of a vine disease, is the type species. In 1911 Beauverie (1914) studied the fungus of muscardine, comparing it with yet another closely related species (*B. effusa*) found on silkworms, and he showed that the two species possessed common properties and that it was necessary to create a new group to include them. Shortly thereafter (1912) Vuillemin, who in 1910 was preoccupied with modifying the classification of Hyphomycetes of the Fungi Imperfecti, created the genus *Beauveria* of which the species *bassiana* became the type. Although Clements and Shear (1931) have since placed the genus *Beauveria* in synonymy with *Phymatotrichum*, most authorities retain the former name. Therefore, according to present usage, the name of the fungus responsible for muscardine of the silkworm is *Beauveria bassiana* (Balsamo) Vuillemin. Following the original taxonomic work on *B. bassiana*, a number of workers made comparative studies of the several species in this genus, including *Beauveria effusa*, *B. densa* (Lk.) (= *B. tenella* Del.), and *B. globulifera* (Speg.), all of which are capable of infecting the silkworm. The distinct identity of the genus is now generally recognized.

The morphological and physiological characteristics of *Beauveria bassiana* have been studied in culture as well as in nature. The fungus grows well, at an optimum temperature of approximately 28°C., on most of the artificial media used to cultivate fungi. It characteristically produces a flat, mealy, chalky, pulverulent growth, with spore formation taking place in from 3 to 7 days. When the spore is placed in water it becomes swollen in 24 to 48 hours and puts out one or more slender thin-walled germ tubes. About 32 hours later the germ tubes range from a few microns to about 80 microns in length, and the branches arising from these are very short. Conidial development begins at this stage, and later, when conidia are abundantly produced, they are borne in rather compact globose heads, either on the main hyphal branches or on short laterals that are usually at right angles to the main axis. This branching may be repeated, forming compact heads.

*The Disease in the Silkworm.* Infection of the insect by *Beauveria bassiana* begins soon after the animal's integument becomes contaminated with the spores of the fungus. Certain workers believe that the insect also becomes infected by way of the tracheal openings or the digestive tract. Infection by these routes, however, apparently occurs infrequently. It appears that in most instances the infection occurs by a direct penetration of the integument by the infecting germ tube. The germ tube is produced within about 2 days after the spore or conidium lands on the insect. Infection is facilitated by the presence of warm temperatures and humid atmosphere; although if these conditions are maintained for more than 24 hours, the rate of infection is usually retarded. As the mycelial filament penetrates the chitin the latter appears to be dissolved or digested. The filament penetrates through the cuticle and is met by an increased number of blood cells which become intermingled with the fungous threads. Phagocytosis may take place to some extent, but apparently it does not afford much real protection, since the fungus usually keeps on developing even at the expense of the blood cells.

During this invasion the hypodermis is destroyed in the area immediately surrounding the infecting hypha. Hypodermal cells adjacent to this area also show pathological manifestations. For example, large vacuoles appear in the distal ends of the cells, i.e., the end next to the newly forming cuticle. Furthermore these cells do not stain so deeply with acid fuchsin as do normal cells.

Once within the body cavity of the silkworm, the fungus continues its development in a characteristic manner. As early as 1853 Vittadini gave a fairly accurate description of this process. He showed that in the blood of the insect the fungus multiplies in the form of short filaments or hyphal bodies. The volume of the blood diminishes, and the blood cells are destroyed in proportion to the development of the disease. The physical-chemical properties of the blood change, as is indicated by the fact that the acidity diminishes and approaches neutrality. The formation of unidentified crystals in the blood has also been reported. As the infection proceeds, the circulation is slowed, then stops, and the consistency of the body becomes pasty. General paralysis sets in, followed by the death of the insect. According to Paillot (1930), invasion of most of the caterpillar's solid tissues, such as the fat body, does not occur until after the death of the insect.

After death the insect's body takes on a reddish tinge and becomes more and more hardened. The exact nature of this red coloration has been a subject for argument. Perroncito as well as Masera believe it to be due to the presence of the red-pigmented bacterium. *Scrratia marcescens* Biz. Within 24 to 48 hours after death, the body is covered with a white network, which finally takes on a mealy aspect after the formation of the conidia. At the same time there appears a white inflorescence of an essentially crystalline nature. Verson gave the composition of this material as a double oxalate of magnesium and ammonium. Its true nature and identity have not been confirmed. A common post-mortem change seen in some insects infected with *B. bassiana* is the liquefaction of the internal tissues. This liquefaction is usually not followed by the formation of spores by the fungus.

That the lethal action of *B. bassiana* may not be entirely parasitic is indicated by Dresner's (1947) observation that a steam-acetone extraction of mycelium produces a substance that, even when greatly diluted, has a marked insecticidal action against certain mosquito larvae. He also noted that the germinating fungus secretes a chemical that possesses a knockdown effect on houseflies. Within half an hour after being dusted with spores of the fungus, the flies began dropping from the walls and ceiling of the moist chamber in which the tests were run. Within 3 hours there was 100 per cent knockdown with no subsequent recovery. A very high relative humidity appeared to be necessary to bring about the phenomenon.

Individual silkworms show very little true immunity against the fungus. The fact that larvae of the fifth instar are more sensitive to infection than are those of the first instar does not in itself represent a true immunity in the latter but rather is the result of mechanical and physiological barriers. Apparently there has been a diminution in the severity of the disease since the early part of the nineteenth century. Some workers have postulated that this may be due to a decrease in the virulence of the fungus, a developed immunity in the insects, or both. There is not proof that any of these occurs, however, and it is more likely that the decrease in incidence has been the result of better handling of the silkworms and the adoption of more sanitary precautions. Strains of *Beauveria bassiana* kept on artificial media have been known to retain their virulence for extremely long periods of time. Frequent transfer does lower its virulence somewhat, and it is therefore better to keep a single culture for a long period of time than to transfer it frequently if the maintenance of a high virulence is desired.

*Transmission of Fungus. Beauveria bassiana* is transmitted from one silkworm to another principally through contact and contamination. The fact that infection occurs by the cutaneous route enhances the effectiveness of this mode of transmission. The fungus is not transmitted through the egg. Adult moths, however, may contract the disease, and, in fact, death may occur more rapidly among them than among the caterpillars. Although not of much consequence, transmission from one generation to the next may occur when conidia contaminate the exterior of the eggs and thus infect the larvae that hatch from them. Such contaminated eggs may be disinfected and the danger of infection thus removed.

The principal means by which the fungus is transmitted from one generation of insects to the next is made possible by the great longevity of the conidia or spores. These remain alive from one season to the next in the rearing rooms, and it is by such preservation that the disease is carried from generation to generation in most cases. Although in moist air the duration of vitality would be considerably shorter, it has been shown that the spores remain viable in dry air for at least 5 years.

Methods of combating Muscardine of the Silkworm. Measures employed in the control of silkworm muscardine may be divided into two groups: those taken during the course of rearing and those taken between rearing seasons or without regard to these seasons.

When the disease strikes during the rearing season, the sick catepillars should be removed and burned before the appearance of the conidia, which are formed about 48 hours after the death of the insect. The ailing larvae may be picked out by hand, but a more convenient method is that of using a wire mesh or perforated paper as a barrier. The healthy caterpillars have the ability and energy to crawl through the openings in this barrier to fresh food; the diseased ones remain behind and with the litter should be collected and burned.

Numerous attempts have been made to apply chemical and gaseous disinfectants without interfering with the vitality and development of the silkworms. Formaldehyde vapors and burning sulfur have been used in Europe with varying degrees of success. They are somewhat impractical because of the care necessary to maintain a dosage which is effective and yet which is not harmful to the insects. For example, the burning of 100 grams of sulfur per cubic meter is effective, but it also is injurious to the insects. Smaller dosage, 20 to 30 grams per 100 cubit meters, repeated daily for considerable periods has been recommended. Similar difficulty has been had with the use of other compounds. Masera (1940) found "Procid", a mixture of copper oxychloride and hygroscopic substances, to be very effective against the fungus, but toxic against larvae that ate it. Similar results were obtained with sodium bisulfite.

Between rearing seasons, the equipment and surroundings should be thoroughly cleaned and disinfected. Almost any effective fungicide may be used to destroy the fungi, as long as it is capable of destroying the resistant spores. For many years fumigation with sulfur in the presence of water vapor was used, especially where expense was a factor. Form aldehyde is perhaps one of the most effective vapors, since it is penetrating as well as fungicidal. Commercial formalin is useful in disinfecting the rearing equipment, as are numerous manufactured fungicides. A commonly recommended disinfectant is copper sulfate used in a 5 per cent solution. Some writers suggest a much lower dilution (e.g., 1 pound of copper sulfate in 100 gallons of water). Pulverized copper sulfate powder added to walks and pathways leading to the nurseries is another precautionary measure.

In countries like France and Italy, where the disease has been particularly severe, laws and regulations have been established to aid in its control. When the disease makes its appearance in a silkworm nursery it must be reported to the proper officials. Quarantine placards are posted, and no one associated with a diseased colony is permitted to visit a healthy one.

The measures mentioned here for the control of silkworm muscardine may, to a considerable extent, be applied to most fungous diseases of other insects. Modern insectaries, as well as individual entomologists, are occasionally troubled by the occurrence of fungous diseases among stocks of insects being reared for experimental and other purposes. The experiences of sericulturists with muscardine provide information that may have general application.

**Beauveria bassiana Infections in Other Insects**

In North America, at least 30 species of insects have been reported as hosts of *Beauveria bassiana*. Numerous additional species have been cited from Europe and other parts of the world. Some of these insects represent serious agricultural pests, and their susceptibility to this fungus deserves further mention.

*Infection in the Corn Borer*. The European corn borer, *Pyrausta nubilalis* (Hbn.), is readily attacked and killed by *Beauveria bassiana*. This was shown with certainty when Metalnikov and Toumanoff (1928) showed that *B. bassiana*, as well as certain other fungi, was capable of causing disease and death in the corn borer when experimentally infected. Corn-borer larvae were much more sensitive to infection than were larvae of the wax month, *Galleria mellonella* (Linn.), which was also susceptible. The first report of *B. bassiana* on the corn borer in North America was that by Lefebvre (1931), who observed the mycosis among laboratory specimens at Arlington, Massachusetts, and imported from Manehuria. He was able to distribute the disease readily both in the laboratory and in the field. In 1934 Barlett and Lefebvre reported on rather extensive field tests using the fungus against the corn borer in eastern United States with encouraging results. A considerable reduction in the larval population was obtained after fields of infested corn and weeds were dusted with a mixture of spores and flour. Canadian workers then undertook similar field trials in Ontario, obtaining controls of 60 to 70 per cent. They emphasized the fact that the time of application of the fungus is of much greater importance than is the rate of application. A similar emphasis was made by Steyaert (1935) in the case of *Stephanoperes hampei* Ferr., a pest of coffee in the Belgian Congo. In this instance infection with *B. bassiana* must be effected before the insect has tunneled into the berry.

The pathogenesis of the infection in the corn borer has been worked out by Lefebvre (1934). The germinating spore on the surface of the larva produces an infection hypha, which penetrates the integument of the insect at any point except the head. This penetration appears to be assisted by the production, on the part of the hypha, of substances that dissolve the chitinous layer of the body wall and by mechanical pressure exerted by the fungus. Lefebvre asserts that he found definite indications that infection may also take place by way of the alimentary tract. After infection through either point of entry, the fat body is the first structure to be attacked. (The reader will recall that according to Paillot this tissue was not invaded, in the case of the silkworm, until after the death of the insect). The glandular structures and ganglia are next affected, while the last tissues to be attacked are the nervous system, the gonads, and the muscles. The fungus frequently penetrates the tracheae, greatly impairing the process of respiration.

The external appearance of a diseased corn borer is similar to that of a silkworm infected with the same fungus. Such a larva first becomes sluggish in movement, fails to respond to most external stimuli, and in most instances turns a characteristic pink color. The larva remains soft and pliable until the mycelium has ramified and grown throughout the various parts of its body. Following this the body becomes rigid and mummified. It can now be readily broken, and the whitened chalky contents can be crumbled into a power. No external signs of the fungus are evident as long as the mummified larva is kept in a dry atmosphere. Soon after exposure to moist air, however, the white mycelium becomes apparent over the surface of the insect. Within a day or two conidia are produced abundantly giving the insect a mealy, powdery appearance.

Certain phases in the course of infection taken by entomogenous fungi in the corn borer have also been given attention by Toumanoff (1933), who worked not only with *Beauveria bassiana* but also with *B. globulifera* (Speg.), *Aspergillus flavus* Link, and *Spicaria farinosa* (Fron.). Larvae infected with either of the last two species more slowly or make but slight convulsive movements when touched. Small black spots appear on the skin; filaments of the fungus are found under these spots in a kind of abscess under the destroyed hypodermis. The filaments ramify into the body wall and into the abdominal cavity, eventually destroying all the tissues.

*Infections in Other Insects.* A number of other economically important insects are known to be susceptible to *Beauveria bassiana*. For example, Jaynes and Marucci (1947) found considerable numbers of codling-moth larvae, *Carpocapsa pomonella* (Linn.), dead of a *B. bassiana* infection in New Jersey apple orchards. Laboratory tests showed the fungus to be highly pathogenic when spores of the organism were either dusted upon or inoculated into healthy hibernating or freshly spun-up summer larvae. In the field, artificial dissemination of spores on the foliage brought about a significant increase in mortality. Spore dust or spray applied to infested fruit prevented the newly hatched larvae from making a successful entrance and also killed the larvae after entry. To a certain degree, therefore, the fungus is an agent of natural control on all stages of the larval population. It appears to be particularly important in cool wet seasons. Infected codling-moth larvae are characterized by the dark-brown lesions over the epidermis and a weakened condition of the insect, which becomes typically mummified after the penetrating hyphae have consumed and displaced the body tissues. The fungus is capable of killing the larvae even when the moisture content of the air is low, but the external mycelial growth with its consequent sporulation is not produced unless considerable amounts of moisture are present. The fungi within the mummified larvae will remain viable for months and give rise to the white mycelial growth when moisture is supplied. Jaynes and Marucci assert that in a few infectivity tests larvae showing only a few lesions were able to recover. Since such recovery is rarely seen in insects, an investigation into its nature appears warranted.

Although not observed in large-scale epizootics, *B. bassiana* has been found pathogenic for such insect pests as the Japanese beetle and the Colorado potato beetle. The latter insect is host to another *Beauveria, B. doryphorae* Poi, & Pat., in Europe were a number of important pests, such as *Pieris brassicae* (Linn.), are hosts to *B. bassiana*. The course of the infection in *Pieris* has been described by Arnaud (1927).

**White Muscardine of the Chinch Bug**

Although at least six species of fungi have been reported as parasitic on the chinch but, *Blissus leucopterus* (Say), only two have been found to be of significance as far as actual epizootics are concerned: *Beauveria globulifera* (Speg.) and *Empusa aphidis* Hoff. The latter is also a common entomophthoraceous parasite of aphids. It is frequently called the "gray fungus" to differentiate it from *beauveria globulifera*, the "white fungus."

The chinch bug, one of the most injurious insect pests of cereal crops in the United States, was first noticed in 1783 in North Carolina. In 1840 it was reported in Illinois, and it has been under observation ever since throughout the central Mississippi Valley states, rendering its greatest damage in the Corn Belt.

Apparently the first observer to record the presence of a white fungus on the chinch bug was Shimer, in 1865, at Mount Carroll, Illinois. Although the identity of the fungus was not determined, Shimer credited it with causing the destruction of vast numbers of the insect, especially in low, creek-bottom land during moist warm weather. His belief that it was a contagious disease did not gain ready acceptance by most entomologists, and several prominent ones scoffed at the idea. Some years later, in 1882; Forbes, in Illinois, and Popenoe, in Kansas, confirmed Shimer's observations when they reported the susceptibility of the chinch bug to fungous attack. The fungus seen by Forbes and Popenoe was the so-called "gray fungus", *Empusa aphidis* (Hoff.).

The white fungus, *Beauveria globulifera* (Speg.), was observed on chinch bugs by Forbes in 1887, in Clinton Country, Illinois. Shortly thereafter it was reported from Minnesota, Iowa, Ohio, and Kansas. Thus more than 100 years elapsed between the time the chinch bug was discovered and the time the insect was found infected with the white fungus.

*The Fungus. Beauveria globulifera* (Speg.) Pic. (= *Sporotrichum globuliferum* Speg.), first described in 1880 from a wireworm adult from Argentina, has been reported on approximately 75 species of insects in the United States, and on numerous additional species in other parts of the world. Its parasitism does not appear to be so specialized as is that of many entomogenous fungi, since its hosts are in numerous families and in several orders of insects. It also grows on corn and certain other plants but not so well as on insects.

On its insect host, *B. globulifera* usually appears as a loose white cottony or mealy growth, at times almost completely enveloping the insect. As short irregular intervals the conidiophores bear minute heads, sessile in attachment and creamy-white in color, composed of conidia closely packed into a nearly spherical form. Some of the characters of this fungus on artificial media have been described by Pettit (1895), Lefebvre (1931), and others. They are essentially the same as those of *B. bassiana* (Bals.).

Since temperature, humidity, and possibly light are variable factors that are able to cause marked changes in the activity of the fungus, a consideration of their effects is warranted. Such a study was made by Headlee and McColloch in 1913; and, because their data might in a general way apply to other muscardine fungi, it is quoted from their publication:

*Temperature.* Field temperatures may influence the fungus in two ways—the extremes may cripple and destroy it, and the optimum encourage and further its growth.

Careful laboratory studies at this station have shown that the exposure of spores of the fungus to 104°F. to 105°F. in a saturated atmosphere for twenty-four hours does not prevent strong growth, that an exposure for forty-eight hours destroys most of the spores, and that an exposure for seventy hours does not kill them all. Spores allowed to develop for forty-eight hours, and then exposed to this temperature and moisture, perish. Spores previous to germination, in a comparatively dry atmosphere, may be exposed for five hours to as high as 209°F. without injuring the germination.

Spores, dry or wet, may be exposed over night to low temperature, even when the mercury reaches as low as $-18°F.$, without apparent injury. Spores freshly sown were exposed to changing temperatures from January 5 to February 12, during which time the temperature changed from freezing to thawing, or *vice versa*, twenty-one times—the daily mean was below 32°F. for nine days, and the minimum temperature 15°F.—without injury. Growing fungus threads subjected to repeated freezing and thawing temperatures are not destroyed.

While these facts render it quite possible that the fungus growth, encouraged by moist, cool weather in exposed and cultivated fields, might be destroyed by succeeding hot weather, for the exposed dusty soil will reach about 135°F. at the surface, it is very unlikely that the ordinary foci of the disease, sheltered as they are beneath heavy growth of weeds and grass, would be even seriously injured, for in such location the temperature rarely or never reaches 91°F. It is not likely that the fungus, even in cultivated fields, would be eradicated by dry, hot weather, because while exposed mycelium would perish, the dry spores can withstand far higher temperatures than they would experience, and in the process of cultivation much of the fungus must be thrown beneath the surface of the ground, where the temperature rarely exceeds 100°F.

In view of the fact that the maximum air temperature of the last twenty-five-year period at Manhattan has been 113°F. and the minimum − 35°F., it does not seem probable that extremes of temperature such as the fungus is likely to experience in Kansas will ever seriously reduce, not to say eradicate it.

On bee-broth agar the fungus makes good growth at 70°F and 80°F., less vigorous growth at 90°F., and no growth at all at 50°F. and 100°F. The growth comes a little more quickly at 80°F. than at 70°F. The optimum temperature for its growth and spore production on this medium is therefore between 70°F. and 80°F., probably about 75°F.

As might be expected, the temperature for best growth on chinch bugs is the temperature for best growth on this medium. Only three temperatures were given careful trial—50°F., 70°F., and 90°F. The bugs succumbed to the disease most readily in 70°F.

In general, the average mean temperature in this state [Kansas] is such that the fungus can grow well from April to October, while it is more or less completely dormant from October to April.

*Moisture*. Moisture of two sorts appears to influence the growth of the chinch-bug fungus—relative humidity and water.

The fungus will not grow in a relative humidity of 90 per cent or less, but will remain dormant for an indefinite period. We have kept it in dried corn-meal culture for more than eighteen months, and found it perfectly virile at the end of that time; found it would grow readily and would destroy chinch bugs. It is hardly conceivable that the chinch-bug fungus could have too much moisture in a state of nature. Dashing and washing rains might carry much of it away, but would serve merely to distribute it, and enough would in all probability be left to carry the disease on. The best degree of moisture for germination seems to be a film of water, although germination will take place in a relative humidity of a little less than 100 per cent.

These facts indicated that while the fungus can not grow in dry weather, it is unlikely to be eradicated by any extreme of moisture it is likely to experience in Kansas. As a matter of fact, this agrees with universal experience with this fungus. It thrives in wet weather, disappears during dry periods, and springs up again on the advent of sufficient moisture.

*Light*. General germination of the spores in a film of water required under a temperature of 75°F., twelve days, while subjected to normal daylight coming through north windows, and when in complete darkness germination became general in one day less. In a confirmatory experiment general germination was made simultaneously in thirteen days. Ordinary daylight is thus shown not to be seriously hostile. This also is borne out by field experience.

*The Infection in Chinch Bugs*. The course of the infection with *Beauveria globulifera* in the chinch bug is very much like that which takes place with similar fungi in other insects. Under very humid conditions, the conidium lying on the integument of the chinch bug germinates by sending out a tube that penetrates the body wall of the insect. The bug dies of the infection in about 3 days. Hyphae fill the body cavity of the bug, finally penetrating to the outside, where the body is covered with the typical white mycelial growth. Numerous conidia are formed on the conidiophores, and when these land on other insects the cycle is repeated if conditions of temperature and humidity are favorable. As might be expected, the disease spreads most rapidly when the bugs are crowded together or are very numerous on the host plants.

According to Smith (1933), the fungus also kills chinch bugs, beetles, and other insects during the winter and in the early spring while they are still in hibernating quarters, in grass clumps, under stones, boards, and in wheat fields. The fungus thus can overwinter in protected situations that supply the proper moisture conditions, and it is carried to the field by the spring migration of the chinch bugs.

As demonstrated by Snow (1896), healthy chinch bugs may be infected with *B. globulifera* experimentally. This may be accomplished simply by dusting the insects with the fungous spores. Chinch bugs of all ages are attacked, but the older the bug the more easily it succumbs. Bugs that have been weakened or injured from any cause appear to be more susceptible to attack than are those which are in perfect condition. Adults of the overwintering second brood are more resistant to the fungus than are the adults of the first brood.

*Use as Control Agent*. The efforts of early workers to use *B. globulifera* in the control of the chinch bug constitute an interesting and important phase in the history of microbial control. The failure of the fungus to fulfill the high hopes of those who advocated its use serves as a warning to be remembered whenever excessive enthusiasm is expressed over the control value of newly discovered insect pathogens, especially when this control depends upon the artificial distribution of the microorganism.

The first attempt to bring about an outbreak of the disease by artificial dissemination was that by Lugger, in 1888, who distributed diseased bugs in several localities in Minnesota. Although the experiment appeared successful, Lugger suspected that, since the disease spread so rapidly, the spores of the fungus were already present in the test fields and that he had only reintroduced them.

In 1888 F.H. Snow began his observation and experiments in Kansas; these extended through 1896. In 1891 the Kansas state legislature established an experiment station at the University of Kansas to propagate the fungus and to distribute it free of charge. It was placed under Snow's direction. Almost 50,000 packages of the fungus were distributed by this organization, but the true value of the program was never ascertained with certainty. The reports of observers in 1891 and 1892 were very favorable, whereas those during succeeding years were less favorable. The probability that the fungus was widely distributed naturally could not be ruled out.

Distribution programs were carried on in states other than Kansas, but in each case the work was eventually dropped. Lugger in Minnesota tried it again in 1895 but abandoned the method by 1902. It was similarly abandoned in Illinois, Nebraska, Missouri, Ohio, and Oklahoma. The artificial distribution of the fungus did not appear to affect materially the incidence of the disease in chinch-bug populations or the effectiveness of the control brought about by the infection.

An appraisal of the effectiveness of these distribution programs was made easier by the work of Billings and Glenn, who, in 1911, published the results of their comprehensive observations on the disease. The following points were among those set forth in their conclusions.

1. The chinch-bug fungus is present naturally in fields everywhere throughout the infested area in Kansas.

2. It is present in such great abundance that any artificial distribution of infection in a field would be too insignificant, by comparison, to be of practical use.

3. Its distribution naturally through a field is much more uniform than any artificial distribution can be made.

4. The amount of fungus used experimentally in both wheat and corn fields was so far in excess of any that would be used by the farmer in infecting his own fields that he could not reasonably expect to succeed.

5. The fungus shows little tendency to spread from centers of artificial infection. The apparent rapid spread of the fungus is due to favorable conditions bringing it into activity simultaneously over considerable stretches of territory.

6. In fields where the nature presence of the fungus is plainly evident its effect on the bugs cannot be accelerated to any appreciable degree by the artificial introduction of spores.

7. In fields where the fungus is not in evidence spores introduced artificially have no measurable effect.

8. Apparent absence of fungus among chinch bugs is a field is evidence of unfavorable conditions rather than lack of fungus spores.

9. All the benefits of the Sporotrichum [Beauveria] disease of chinch bugs may be realized by merely letting the fungus naturally present in the soil do the work of extermination as far as it will.

10. Moisture conditions have much to do with the appearance of chinch-bug disease in a field; artificial infection nothing.

11. Spent adult chinch bugs succumb to attack more readily than younger ones, but as the old bugs have finished depositing their eggs, their loss by fungous disease accomplishes little else than increasing the amount of the infectious material.

12. Laboratory experiments can be made to prove that artificial infection accomplishes results upon bugs confined in cramped quarters and without food, but in the field, where fresh and usually drier air prevails and food is abundant, an entirely different situation is presented.

13. Advocating artificial infection or encouraging it by sending out diseased chinch bugs does not serve the best interests of the farmer, since his attention is thus diverted from other and more efficient methods of combating the pests.

14. The reported success of former years on the part of farmers is believed to be due to the following causes: (1) failure to recognize spontaneous outbreaks of the disease because of previous artificial sowing of infection, and also failure to use check, or untreated fields as a basis of comparison, thus claiming the outbreak as directly due to artificial infection; (2) failure to distinguish the skins of molted bugs from dead bugs; (3) mistaking the scattering of chinch bugs in cornfields for evidence of their death by fungous disease when carcasses were not present as proof.

Perhaps one of the most significant conclusions arrived at by Billings and Glenn was that in fields were the fungus was naturally present, its effect on the bugs could not be accelerated appreciably by the artificial introduction of spores. Fawcett (1944) has accounted for this fact by the suggestion that the wind-borne spores are produced in such great profusion, providing wide distribution in a short time, that the saturation point for infection is readily attained and that artificial distribution is therefore of little account. It should be remembered, however, that the conclusions of Billings and Glenn apply particularly to the chinch bug, to *B. globulifera*, and to the localities in which they worked. It would be unsafe to make sweeping generalization regarding all insect mycoses solely on the basis of their findings.

What may be accepted as the most general currently accepted opinion with regard to the control value of the fungus against the chinch bug has been expressed by Packard and Benton (1937). These authorities state that *B. globulifera* is probably the most destructive natural enemy of the chinch bug; that it is generally present in fields throughout the country but that its effectiveness depends upon the weather; and that, since it has been proved that the spores of the fungus are present wherever the bugs are common its artificial dissemination as a control measure is needless.

**Green Muscardine**

Upon examining containers of soil in which had been placed some larvae of the wheat cockchafer, *Anisoplia austriaca* Hbst., Metchnikoff, in 1879, observed some of the specimens to be dead. Two days after the insects died a white fungous growth appeared on their bodies, being particularly noticeable around the spiracles. This growth soon covered the animals entirely except the head. Later it took on a green color, eventually becoming a dark blackish-green. In the blood of the insects oval fungous bodies could be seen, which, as the infection progressed filled the body cavity as a mycelial mass. For this fungus Metchnikoff first proposed the name *Entomophthora anisopliae*. Later, at the suggestion of the botanist Cienkowsky, he called it *Isaria destructor*. Delacroix (1893) published on the fungus under the name *Oöspora destructor*. A few other synonyms (e.g., *Metarrhizium cicadinum* v. Höh.) are to be found in the literature. Although one of the Fungi Imperfecti and not an entomophthoraceous fungus, the original specific name remains valid, and the organism is now known as *Metarrhizium anisopliae* (Metch.) Sor., as it was designated by Sorokin (1879). It is commonly referred to as the "green-muscardine fungus", being characterized by the dark-green color of its conidia, or spores. The green color, however, does not appear to be characteristic of all species of the genus, since another species, *Metarrihizium album* Petch, parasitic on leafhoppers in Ceylon, is white in color, and *Metarrihizium brunneum* Petch, parasitic on a cicadellid in the Philippine Islands, is yellow to brown. On the other hand, a fourth species, *Metarrihizium glutinosum* Pope, isolated from deteriorated cotton, has dusky olive green to olivaceous black conidia.

Metchnikoff's 1879 report is of particular significance, since in an addendum to his article he describes how he experimentally initiated the disease in healthy larvae. It was this experiment that may be credited with establishing the idea of intentionally causing a disease in insect pests of economic importance. Metchnikoff mixed spores of the fungus with the soil and placed the healthy larvae therein. After 10 days, 8 out of 9 larvae were dead of the fungus. In 1880 he reported a similar experiment in which larvae and adults of the sugar-beet curculio, *Cleonus punctinentris* Germ., exposed to spores of the fungus died within 12 days.

Since Metchnikoff's early observations a large number of additional insect species have been reported as hosts of the green-muscardine fungus. This include approximately 70 species in North America, with probably an equal or greater number from other parts of the world. Apparently the first record of the fungus in the United States is that of Pettit (1895), who found it parasitic on the wheat wireworm, *Agriotes mancus* (Say) in New York State.

*Characteristics of the Fungus*. The green-muscardine fungus *Metarrhizium anisopliae* is usually considered to have a systematic position near *Penicillium* in the family Moniliaceae. Its perfect stage is not known, and the report that it is a *Cordyceps* is generally discounted. Pettit (1895) describes a variety *americana* and Johnson (1915) records forma *major* and forma *minor* occurring on different hosts. Most of the variations that have been noticed are concerned with differences in spore sizes and in coloration.

*Metarrhizium anisopliae* grows readily on artificial media and develops particularly well on potato media. Ordinarily it is easiest to start a culture by using the spores of the fungus. With proper techniques, however, cultures may be initiated with nonfruiting forms. Upon germination the spore swells somewhat and a germ tube appears at one end and sometimes at both ends of the spore. As described by Speare (1912), this germ tube elongates, and branching occurs after about 30 hours and continues for several days. Certain of the branches extend upward and become the sporophores. The latter are usually closely packed with their branches intertwined. After about a week they begin to form spores which are at first white, but gradually, as they mature, appear olive green in color. The spores become partly cut off from the grouped sporophores and cohere in prismatic masses. Spore formation occurs by the simple abstriction of elongated buds which appear at the distal ends of the branched sporophores. Other buds form below the original one, and this continues until chains of spores are formed. So closely do the chains of spores arising from adjacent sporophores cohere that in old cultures they may form a crust over the surface of the medium and may be removed in large flakes. The sporophores are short and are branched in a manner somewhat resembling the branching in *Penicillium*, although the long unbranched basal portion of the sporophores usually associated with this genus is not seen in *Metarrhizium*. The spores may vary considerably in size but are usually within a range of from 5 to 7.5 microns in length to 2.3 to 3.7 microns in width.

Although only a meager amount of information is available relative to the physiological properties of *M. anisopliae*, it may be assumed that they are in general similar to those of the white-muscardine fungi already described. High humidities and warm temperatures promote the growth and development of the fungus. In culture, a temperature between 24 and 26°C. appears to be optimum, the entire range for normal development being between 10 and 30°C. The ability of the spores to germinate is destroyed at temperatures between 55 and 60°C. for 5 minutes. The range of hydrogen-ion concentrations for the normal growth of the fungus is from pH 4.7 to 10. The optimum value lies between pH 6.9 and 7.4. Although a medium containing organic matter is more favorable for the growth of *m. anisopliae* than is one having an inorganic nitrogen source, both types of medium can support its growth. No special requirements are necessary as far as the organism's carbon source is concerned. The growth and fructification of the fungus are retarded by rays of the sun. The spores may be kept in a dry condition for 3 years or more.

*The Infection in Insects.* From the knowledge already gained during studies on white muscardine of the silkworm, a general idea of the course of such infections in insects was at hand from the time *Metarrhizium anisopliae* was discovered. Not until 1926, however, was much attention paid to the details of the pathogenesis of *M. anisopliae* infections in insects. In that year Glaser reported on his studies of the green-muscardine disease in silkworms. This was followed 3 years later by similarly detailed reports, by other authors, of the infection as it occurs in the European corn borer.

The silkworm, *Bombyx mori* (Linn.), acquires the infection via penetration of the integument by the germinating hypha. It does not appear to be susceptible via the digestive tract. After the fungus enters the body cavity it develops in the blood and does not penetrated the tissues until after the death of the insect. A reaction on the part of the host occurs, as is indicated by large masses of phagocytes that congregate around and between the fungous threads. The phagocytes are unable to offer very much protection, however, since the parasite is too large for ingestion. Four or five days after infection the larvae begin to lose thier appetites, assume a yellowish color, and appear sick. In 5 or 6 days they become extremely feeble and die, after which they become rigid in form and brownish-yellow in color. The fungus now invades nearly all body tissues.

After the silkworm dies, certain of the hyphae penetrate outward, and 7 or 8 days after the beginning of the infection of hyphae break through the integument everywhere. The larva soon becomes covered with a white down consisting of hyphae destined shortly to become the sporophores. A day or two later the rather short and branched sporophores begin to form spores. These soon cover almost the entire body of the silkworm, and as they mature, the spores become olive green in color, cohering in masses which may be removed in large flakes. The silkworm usually dies in the larval stage. Sometimes, specially among those infected during the last instar, some individuals will spin a cocoon and succumb within. In such cases the white down appears, but the fungus rarely fruits.

In 1929 Wallengren and Johansson published a description of the manner in which *M. anisopliae* infects the European corn borer, *Pyrausta nubilalis* (Hbn.), and in so doing added considerably to the information supplied by Glaser in connection with his silkworm studies. Infection of the corn borer takes place without any great difficulty; the conidia fasten very easily to the intersegmental folds and from there penetrate the integument. The intersegmental folds have a thinner chitinous layer than the other parts of the segments and therefore offer a more favorable place for infection. The germinating hyphae, however, can pierce the fairly thick integument on the dorsal side of the segments.

Upon examining under the microscope a piece of integument or skin dissected from a corn-borer larva, one can see that the areas immediately surrounding the bases of the hairs are darker and brownish and present a granular appearance produced by a number of small, round, dark pigment bodies, gathered close to each other immediately under the surface of the chitin. The rest of the skin has similar granules farther apart. Histologically, the body wall shows a lamellose structure, the lamellae lying parallel to the surface. The black pigmentary bodies lie immediately under the outer heavily sclerotized surface membrane. From each pigment granule emanate one or more threads, clustered together, which continue down into the cuticle to about a third of its thickness. The threads converge so that the entire formation resembles an inverted pyramid. In the areas around the bases of the hairs, however, the threads are not gathered in pyramids but run straight down into the cuticle; they appear to be somewhat shorter than those in the pyramids. With this histological picture in mind, let us now consider the details of infection as conceived by Wallengren and Johansson.

With the proper conditions of temperature and humidity, the fungous spores, or conidia, fastened on the surface of the integument send out germinating hyphae. At those places where the conidia germinate the cuticle assumes a yellowish color. Under this yellow spot the pigmentary bodies become clotted and fall to pieces, and the threads disintegrate. Each hypha pierces the cuticle and begins its entrance into the body wall at the point of one of the thread pyramids, which disintegrates and becomes granular. The outer layer of the cuticle is more or less destroyed, and the area concerned assumes a more yellowish or dark-brown color. These colored spots increase in size as the infection progresses. It is possible for the infection to occur without causing the brown spots. Also, after the hypha has penetrated the skin, the fungus appears to be able to continue its growth without further discoloring of the integument.

The growing conidial hyphae apparently secrete some sort of chitindissolving substance, possible an enzyme, which opens or facilitates the way for penetration. The hyphae also exert a rather great pressure on the substratum. In sections, the cuticular lamellae can be seen to be pressed down when the hyphae are developing at right angles to it. The hyphae often grow for relatively long distances between the lamellae, evidently because growth in this direction offers the least resistance, but sooner or later they turn inward, perforate the endocuticula and hypodermis, and enter the body cavity. According to Wallengren and Johansson, the hyphae seem to have a special preference for the adipose tissue, but they also invade the muscular and central nervous systems. It is not entirely clear, however, whether these workers refer to premortal or postmortal conditions. Since their description at this point apparently refers to living insects, this observation is in conflict with Glaser's assertion that in the silkworm the fungus does not enter the internal tissues until after the death of the insect. In any case, after the corn borer dies, rapid growth of the fungus takes place, filling the interior of the body. Hyphae force their way through the body wall and appear as a white fluff on the surface of the integument. Finally, the characteristically green conidia are formed. Conidia are also formed within the body of the insect.

The symptomatology of the mycosis in the corn borer is similar to that which characterizes the infection in most larvae. The first symptoms include the appearance of yellow or brown spots on the cuticle; these usually spread over other parts of the body. The fungus seems to have some toxic influence on the host. Marked nervous disturbances become evident in the later stages of the disease. The larvae lose their appetites, become apathetic, make no spontaneous movements, and their irritability decreases more and more. Finally the insects lose their righting reflex, and death soon follows. The body finally becomes stiff and mummified. Under conditions of an atmosphere saturated with moisture and a temperature of about 25°C., death occurs in approximately 4 days. Occasionally an infected larva, under conditions not entirely optimum, will survive for a considerably longer period of time. It is unknown for larvae to survive longer than 26 days, however, once the fungus has penetrated the body wall. In corn-borer larvae the mycosis produced by *M. anisopliae* under experimental conditions has a death rate of 100 per cent.

*Use as Control Agent*. Following Metchnikoff's (1880) discovery that the wheat cockchafer, *Anisoplia austriaca* Hbst., and the sugar-beet curculio, *Cleonus punctiventris* Germ., could be artificially infected with *M. anisopliae*, other workers soon considered using the fungus as a control agent on a wide scale. In southern Russia, entomological commissions were established in Kharkov and Odessa to investigate the practicability of artificially distributing the fungus to combat these insects, particularly the cockchafer. In 1886 and 1888, Krassilstschik undertook to use the fungus for combating the sugar-beet curculio, reporting field mortalities of from 50 to 80 per cent. In his laboratory at Smela, near Kiev, he produced spores of the fungus in large enough quantities to make field distribution practical, although still on a rather limited scale.

Further attempts to use *M. anisopliae* against various insect pests were made during the early years of the twentieth century in such parts of the world as Java, Hawaii, Trinidad, and Puerto Rico. In Trinidad, for example, Rorer (1913) used the fungus against the froghopper, *Tomaspis varia*, a pest of sugar cane, with apparently favorable results. The spores were usually distributed in the form of dusts (mixture of flour and spores) applied to the sugar-cane plants at the rate of 2 or 3 pounds to the acre. Specially constructed large culture cabinets were devised to produce the spore material in great quantities.

In some areas, as in Java and in Puerto Rico, the fungus, although an efficient parasite of several noxious insects, did not prove to be an entirely practical means of control. Since it was rather widely distributed in nature, and since its effectiveness depended upon the existence of favorable climatic conditions, many workers decided that attempts to distribute the fungus artificially would be useless. In some localities excellent results were obtained after initial distributions, with a declining effectiveness upon subsequent distributions. The reasons for this are not clear, except that the initial epizootics might have indicated the absence of or the rather low number of naturally existing spores in the area concerned.

In 1913, Friederichs reported excellent results with the fungus when directed against the rhinoceros beetle, *Oryctes rhinoceros* Linn., a pest of coconut. Infected beetles were placed in trap piles of rotten coconut husks and other debris, placed here and there throughout the coconut groves. The beetles gathered in these piles for egg laying. When the larvae hatched, nearly all of them were apparently attacked and killed by the fungus. According to Simmons (1939), however, "no efficient control seems to have come from the introduction" of the fungus into Samoa, although grubs are frequently found killed by the organism, and some officials maintain that it is still a beneficial agency. Simmons points out that regular provision of infected breeding traps is required and that this necessitates constant careful supervision or the results are likely to be the reverse of what is desired.

In Europe a number of attempts were made to use the green-muscardine fungus against the European corn borer with rather encouraging but not convincing results. Hergula (1931), for example, found that, when infested corn plants were dusted experimentally with fungus-spore mixtures, the infestation was reduced to such a point that the damage to the plant by the insect was practically negligible. Data on practical large-scale field application were not gathered.

From the reports and information available, it appears that the value of *M. anisopliae* as a control agent is largely in the realm of natural control. The introduction of the fungus into areas, localities, or islands where it does not occur may be entirely worth while. Once the organism is established, however, it appears that, with certain exceptions, the artificial distribution of spores is useless. If and when favorable weather conditions prevail, the fungus, if present, exercises practically its full effectiveness without the help of man. There seems to be little reason for doubting that the sporadic and small but continuous occurrence of the disease among certain insects has its beneficial effect, even though this effect may rarely be enough to bring the insects under control from an economic standpoint.

**Sorosporella Infection**

Although the infection we are about to consider—one caused by *Sorosporella uvella* (Krass.) Gd.—is sometimes referred to as the "red muscardine", this terminology is not applicable in the same sense in which we have been using it in the foregoing discussions. Unlike the true muscardines, the fruiting bodies of the fungus in this infection are not as a rule produced as a covering on the exterior of the insects; and, instead of becoming mummified and hardened, the insects disintegrate soon after death.

One of the most thorough studies of the fungus and the infection it causes has been performed by Speare (1920). Much of the discussion to follow will be based upon reports of this mycoentomologist.

*The Discovery and Systematic Position of the Fungus.* The fungus under consideration was first observed in Russia by Krassilstschik in 1886. He observed it in the same insect, *Cleonus punctiventris* Germ., on which he did so much work with the green-muscardine fungus, *Metarrhizium anisopliae* (Metch.). Krassilstschik named the newly discovered fungus *Tarichium uvells.* Two years later Sorokin (1888), another Russian scientist, found a fungous parasite of the cutworm *Euxoa segetum* (Schiff.), which he named *Sorosporella agrotidis.* In 1889 Giard pointed out the identity of *Tarichium uvella* and *Sorosporella agrotidis.* Since the genus *Tarichium* was at that time employed for species of *Empusa* and *Entomophthora* in which only resting spores were known, and since the resting spores of the fungus under consideration were unlike those of any known *Empusa* or *Entomophthora*, Giard proposed that the generic name *Sorosporella* be used together with the specific name *Uvella.* Giard's proposal has been generally accepted, even though this workers adhered to the belief that it was an Entomophthorales. That the fungus does not belong to this order was sown by Speare (1920), who considered it a verticilliacious Hyphomycete (Moniliales; Fungi Imperfecti). This remains the systematic position generally accorded it today.

Probable synonyms include *Acremonium cleoni* Wize (1905) and *Fusarium acremoniopsis* Vincens (1915). *Massospora staritizii*

Bresadola (1892) appears almost certainly to be a *Sorosporella*, but since several of its characters are quite different from those of *S. uvella*, it is perhaps another species of this genus.

*Symptoms and Pathology.* Noctuid larvae parasitized by *Sorosporella uvella* exhibit no symptoms of disease during the first few days after being infected. Three or four days before death, however, they lose their appetites and become sluggish in movement. In some host species a change in outward appearance may be observed 1 or 2 days before death. Such larvae may turn a creamy-white color. Sometimes red-colored patches appear a few hours before death, either posteriorly, anteriorly, or more commonly near the middle of the body.

Soon after death, which usually occur within 7 to 10 days after infection, the creamy-white color changes to pink. The pink color appears simultaneously over the entire body, becoming more and more intense until the final development of the fungus is reached and the insect is of a characteristic brick-red color. The body of the insect appears shrunken and wrinkled and somewhat flattened. A longitudinal, ventral, furrowlike depression is nearly always present in its abdomen. The cadaver is soft and pliable, and if a portion of the integument is indented, it remains sunken with little or no reaction. Unlike the case in many other fungous infections, the body is never hard and sclerotiumlike. No fluid exudes if it is pricked with a needle. If the body is torn open entirely, the internal fungus mass is seen to be coherent, of a shiny creamy or pink color, and of a gelatinous consistency.

Later the host's body becomes more shrunken and the reddish color more intense. The body wall is now rather brittle, the slightest shock serving to rupture it. The spores of the fungus become brick red in color, dry, dustlike, and less coherent than previously. Virtually nothing remains of the host's internal organs, and the body appears as a small sac filled with dust.

The histopathology of the disease has been studied to some extent by Speare (1920) in connection with his observations on the yeastlike cells or "blastocysts", which represent the vegetative stage of *S. uvella*. These blastocysts multiply in the hemolymph by yeastlike germination until enormous numbers of them are floating in the circulating blood. They are found throughout the body cavity, within the heart, and wherever it is possible for the blood to penetrate. The hemolymph becomes so loaded with blastocysts that not only is the circulation impeded but its life-sustaining elements seem to become depleted.

The blastocysts apparently never actually invade or penetrate into the living tissues or organs of the host. There appears to be a substance secreted by the fungus which causes the host's tissue to break down. The softer tissues disintegrate first, usually beginning with the fact body. A colony of fungus cells will multiply rapidly near the fat cells; and, after the wall of the fat body has broken down at the surface, the colony will tend to enter and take the form of the organ. As a result, lamellalike or vermiform convoluted colonies are formed. After the fat body, the muscles, nerve fibers, malphighian tubes, alimentary tract, and all the hypodermal tissues are gradually destroyed, until only fragments of the tracheae can be seen. The chitinous body wall, originally very thick, becomes very thin—apparently because of the action of a solvent of some kind.

Phagocytosis of the blastocysts is known to occur. This may consist simply of the engulfment of the parasites within the cytoplasm of individual phagocytic leucocytes, or it may taken the form of phagocytic complexes within which the blastocysts are included; these complexes usually consist of masses of phagocytes arranged in concentric layers about a number of blastocysts. The number of blastocysts contained within a single phagocyte varies from 1 to 15. When large numbers have been phagocytosed, the leucocytes are gradually destroyed. From the information at hand it would appear that the phenomenon of phagocytosis offers little, if any, real protection against the fungus once the infection is under way in a susceptible host. When a small number of *Sorosporella* conidia are inoculated into the body cavity of a relatively nonsusceptible host, such as the silkworm, something occurs, not known to be phagocytosis, which renders the fungus cells innocuous and causes them to disappear.

*Life History of Fungus*. When the blood of a cutworm infected within *Sorosporella uvella* is examined, 6 or 7 days after inoculation, the yeastlike cells, or blastocysts, will be seen floating free in the hemolymph. These blastocysts are the early vegetative stages in the development of the fungus. When young, the cells are elliptical in form, are hyaline, measure 8 by 5 microns in size, and in blood smears occur singly or coherent in pairs, rarely in threes. They multiply by a yeastlike germination; budlike outgrowths appear that grow rapidly until they reach the approximate size and form of their parents, then break away and proceed to form new cells by the same process. Multiplication continues until large numbers of coherent blastocysts are present in the form of colonies throughout the body cavity of the insect. Small colonies coalesce so that large masses are formed which ultimately develop into resting-spore masses. When these spore masses mature they eventually appear typically as brick-red "dust" within the cadaver of the host insect. It is in this condition that specimens of the diseased insects are most often collected in the field. The resting spores, or chlamydospores, mark the termination of the development of the fungus.

In order to determine the origin, nature, and function of the resting spores, Speare (1920) conduced a number of experiments dealing with the organism's life history. Under the stimulus of moisture and suitable temperature, the protoplasm of the resting spores swells, producing budlike protuberances which soon assume the shape of a germ tube, branch freely, and become septate. The fully developed conidiophores are supplied with bottle-shaped branchlets or sterigmata at the tips of which are borne the conidia, which are quite unlike the resting spores from which they arose. The conidia are elliptical in form, thin-walled, vacuolate at each pole, and measure 4 to 6 by 9 to 11 microns in size. The development and structure of the conidiophores and conidia are typical of the verticillate Moniliales, or which the genus *Sorosporella* is a member.

Germination of the resting spores in unbroken larvae may be induced by placing the insects in circumstances that provide the proper moisture and temperature requirements. In a few days an external fungous growth on the unbroken integument of the host becomes visible. The germinations thus obtained are confined only to those spores which lie near the surface, immediately beneath the integument. When, however, the infected larvae are torn open and the red spore masses are freely exposed to the air and light, the germination may involve practically all the spores. The germ tubes from adjacent spores may cohere in such a way that *Isaria*-like fascicles of conidiophores are produced. It appears, therefore, that although the germ tubes are capable of breaking out through the cuticula of the insect, the process of germination is facilitated considerably when the larva becomes disintegrated in the soil.

When the conidia are placed in contact with healthy insects, either externally or internally, infection may be readily induced. In all probability the conidia send out germ tubes that penetrate into the body cavity of the insect where they produce bodies that give rise to the yeastlike blastocysts. Thus, in summarizing what is known of the life cycle of *Sorosporella uvella*, it might be briefly said that the thick-walled resting spores germinate, producing conidiophores upon which are borne thinwalled conidia; the conidia in turn probably send out germ tubes, which, after invading the host insect, give rise to the vegetative cells known as "blastocysts", which multiply rapidly, ultimately becoming resting spores.

The resting spores do not necessarily require a long period of rest before germination, as is the case with certain other fungi. Germination may take place at once if suitable conditions are provided. The spores may be viable after being held in a dried state for 14 months, and they are able to withstand the low temperatures that prevail during the winter months in northeastern United States. Thus the resting spores are able to tide the fungus over periods of unfavorable conditions. The thinwalled conidia, on the other hand, apparently have the function of spreading the organism rapidly while favorable conditions persist.

*S. uvella* can be cultivated on a variety of nonliving media. Although its normal method of vegetative development within infected insects is by means of the yeastlike budding cells, on media a semifilamentous growth is obtained. As in nature, no perfect or ascigerious stage has so far been observed in any of the artificial cultures.

Distribution in Nature. Since its original discovery, *S. uvella* has been reported as a parasite of a number of lepidopterous and coleopterous insects in North America and in Europe. It has been reported from cutworms in Canada, and in the United States at least 10 insect species are known to be susceptible. In the United States it was originally collected in the field on the striped cutworm, *Euxoa tessellata* (Harr.). It has also been found killing the corn earworm, *Heliothis armigera* Hb., in Virginia. Barber and Dicke (1937) report that high-humus-content soil appears to be very favorable to the development of the fungus and that, during a 5-year experimentation period, no moths of the corn earworm emerged from hibernation in this soil. In Florida considerable reductions in populations of mole crickets (*Scapteriscus*) have been reported through the agency *S. uvella* and the green-muscardine fungus, *Metarrhizium anisopliae* (Metch.). It is probably more widespread than is generally recognized; the relative inconspicuousness of the fungus probably accounts for its being overlooked in many cases.

In Moravia the poppy-root weevil, *Stenocarus fuliginosus* Marsh, is occasionally found parasitized by the fungus, as is the sugar-beet curculio, *Cleonus punctiventris* Germ. Some authorities maintain that *S. uvella* is a more effective enemy of *Cleonus* than is *M. anisopliae*. Along with the green-muscardine fungus, *S. uvella* kills a considerable number of the larvae of the black beet weevil, *Psalidium maxillosum* Fabr., in the Russian province of Krasnodar. Most *Sorosporella* infections, however, are found in lepidopterous insects, the noctuids, e.g., *Euxoa segetum* (Schiff.), being particularly susceptible. An accurate estimate as to the probable amount of natural control effected by this fungus has not been made. Since it has been shown that infection of larvae may result by various methods of inoculation (direct contact, feeding, and spray methods), it can be assumed that transmission in nature is probably readily maintained as long as all environmental conditions continue to be satisfactory.

## MYCOSES OF BEES

A number of species of Fungi Imperfecti are known to be capable of parasitizing bees, particularly the honeybee, *Apis mellifera* Linn. That this beneficial insect suffered from mycoses was not generally realized until after the bacterial diseases of brood were recognized. The first comprehensive studies on the fungous diseases of bees were accomplished in Europe. Very little was done in this connection in North America until Burnsude (1930) undertook a study of the fungi associated with the honeybee and their pathogenicity for this insect. Although there have been subsequent investigations and reports on this subject, our information is still meager, and much remains to be learned of the epizootiology of these diseases. One reason for this lack of attention is probably the fact that fungous diseases of bees appear to be less serious and less destructive than the common bacterial diseases of these insects. Furthermore the mycoses rarely become epizootic in intensity, although such outbreaks may occur among wild bees.

In Europe, two species of fungi, *Aspergillus flavus* Link and *Pericystis apis* Maassen, are recognized as the etiological agents of diseases of brood and adult bees. Since *A. flavus* attacks worker brood and adult bees, it is generally considered of greater economic importance than *P. apis*, which attacks only drone brood. When affecting brood the *A. flavus* infection is commonly known as "stone brood" (*Steinbrut*), and the *P. apis* infection as "chalk brood" (*Kalkbrut*). Other fungi that have been reported from Europe as capable of infecting and killing bees and their brood include *Trichoderma lignorum* Tode, *Mucor mucedo* Linn., and the yeast *Saccharomyces apiculatus* Hansen. Although not ordinarily pathogens of live bees, *Penicillium glaucum* Link and *Pericystis alvei* Betts are two of most common fungi in European hives.

In the United States, *aspergillus flarus* attacks brood of the honeybee more frequently than other fungi (Fig 1.4). Its has also been recorded on five or six other species of insects in this country. *Pericystis apis* is not known to occur in North America. In addition to *A. flavus* other forms of *Aspergillus* known to attack bees include *A. fumigatus* Fres., *A. nidulans* (Eid.), *A. niger* Tieg., *A. glaucus* Link, and *A. ochraceus* Wilh. Burnside (1935) found *Mucor hiemalis* Weh. pathogenic for young bees when the latter are exposed to a temperature of about 20°C. By inoculation a 100 per cent mortality may be obtained with this fungus, of which the chlamydospores from infected bees appeared to be the most virulent form. A number of yeasts (*Saccharomyces)* are known to be pathogenic to the honeybee when introduced into its body cavity.

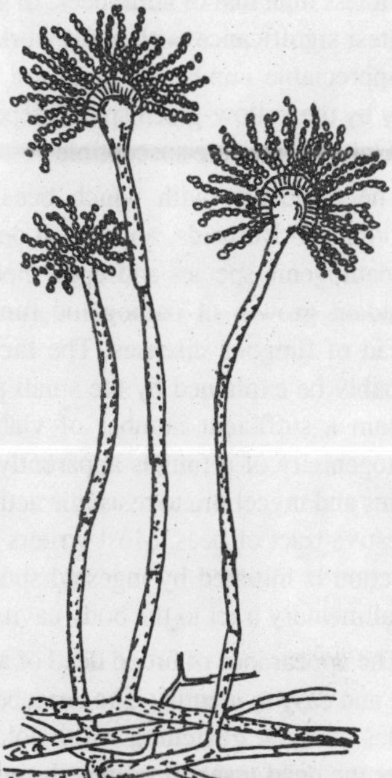

**Fig. 1.4.** *Aspergillus flavus* Thom & Church, which, under certain conditions, may cause a serious mycosis of brood and adult honeybees.

*Nature of the Mycoses*. During his experiments on the patho-genicity of various fungi associated with bees, Burnside (1930) observed several interesting aspects relating to the nature of the mycoses concerned. We shall repeat here the statement of some of his findings.

In the first place, Burnside observed that the greater number of fungous species that cause diseases of adult been also attack the brood. He found this to be the case with most of the species or strains of *Aspergillus* used in his investigations. When inoculated experimentally with a pathogenic *Aspergillus*, brood is attacked and killed more quickly than are adult bees, though the loss of brood resulting from *Aspergillus* mycosis is much less than that of adult bees. In general, mycoses of bees reach their greatest significance with adult workers. Throughout the active season, an appreciable number of these are killed by pathogenic fungi, principally by the yellow-green spored aspergilli. All races of bees common in the United States are susceptible.

The frequency with which bees are attacked by fungi in nature, according to Burnside, appears to depend chiefly upon the virulence of the pathogenic species and upon their dispersion. Conditions that favor abundant growth of pathogenic fungi in nature are conducive to the spread of fungous diseases. The fact that brood is rarely attacked can probably be explained by the small probability that the larval food will contain a sufficient number of viable spores to cause infection. The pathogenicity of a fungus apparently is determined by the ability of its spores and mycelium to resist the action of the intestinal fluids within the digestive tract of bees. Most writers on the subject seem to believe that infection is initiated by ingested spores, the infection progressing from the alimentary tract to the body cavity and internal tissues.

The appearance of brood dead of a fungous disease is fairly characteristic and easy to identify. The larva becomes noticeably harder soon after it dies, and the glistening white color changes to a dull creamy white. Later the dead insect becomes shrunken and wrinkled. The anterior end dries most rapidly, often curving upward at first, later tending to straighten out again. The fungus grows through the body wall, at first just back of the head in the form of a white ring or collar; a day or two later it covers the entire insect. As the fungus continues its development, spores are formed on the external mycelium giving it a green, yellow, black, or other color depending on the particular species of fungus concerned. As the spores and insect carcass become old and dry, the color usually fades.

When adult bees are infected with a fungus they become restless and weak. The weakness increases until death occurs. Some of the infected bees may die in the hive, but usually they fly or crawl away to considerable distances before dying. An increased firmness of the insect's body can sometimes be noticed at the time of death. It is usually more noticeable a few hours later. Under suitable conditions the external production of spores takes place, giving the insect a characteristic mealy appearance.

Since nonpathogenic or saprophytic fungi may develop on the cadaver of a bee dead from another cause, care should be exercised in ascribing the death of the insect to the fungus present, unless the examination is made directly after the death of the bee.

The exact cause of death has not been determined in the case of most of the mycoses, although the destruction to the host's tissue caused simply by the growth and development of the fungus is probably sufficient in most instances. There is evidence, in certain instances at least, that in addition to this physical action there is a chemical or toxic action exerted by the fungus. Undoubtedly a certain amount of digestion of the insect's tissues caused by the activity of fungous enzymes takes place. *Aspergillus flavus* has been shown to produce a transient toxic substance that is capable of causing a fatal poisoning in bees (Toumanoff, 1928; Burnside, 1930). In fact, Toumanoff (1931) believes that the pathogenic action of this fungus is due almost entirely to its toxic action.

*Economic Importance.* The importance of mycoses among bees is difficult to estimate accurately. Larvae may be carried out of the hive soon after they become infected, and adult bees usually die away from the hive. The mortality also varies somewhat according to the species of fungus involved. Some species attack adult bees in nature with considerable frequency; others attack the insects so rarely that under normal conditions their importance is negligible. Within the hive, the death rate may vary according to the condition and inherent strength of the colony. If wintering conditions are poor, allowing moisture to collect in the hive, and if the colony is weak, mycoses are likely to be a more serious threat than otherwise.

In general, it may be said that only slight losses of brood are caused by fungous diseases, and these are of negligible economic significance. Losses of adult bees from fungous infection are also of little economic importance, except that, when pathogenic fungi grow within the hive on combs, frames, dead bees, and the like, late in the winter or early in spring, significant losses are likely to occur. Such losses, however, can be largely remedied by exercising a few simple precautions.

Great care should be taken to ensure the proper ventilation of the hive and to provide for the escape from the hive of metabolic water vapor. Bees dead of fungi or covered with spores should be cleaned from the hive whenever this can be done without exposing the colony to severe weather conditions. All combs and equipment covered with fungi or "mold" should be washed or dipped for a few minutes in 20 per cent formalin, or other effective fungicide.

**Other Fungi-Imperfecti Infections**

The entomogenous Fungi Imperfecti are so numerous that it has been necessary to limit detailed discussion to a few of the better known examples. Furthermore it is impractical even to mention examples of all the remaining genera of imperfect fungi known to include entomogenous species.

Representatives of some of these miscellaneous genera include *Vermicularia cicadina* Ell. & Kell. on cicadas (some authors regard *Vermicularia* and *Colletotrichum* to be of the same generic type), *Monilia penicillioides* Del. on scarabaeids, *Trichoderma viride* Pers. on bees and culicids, *Cylindrodendrum suffultum* Petch on tipulid pupae, *Acrostalagmus aphidum* Oud. and *Cladosporium aphidis* Thuem. on aphids, *Acremoniella verrucosa* Ton., on the clover leaf weevil, *Stemphylium botryosum* Wallr. on coccids, and *Synnematium jonesii* Speare on various Hemiptera.

Approximately six species of *Penicillium* have been reported as being parasitic on North American insects. About the same number of species of *Sporotrichum* have been similarly reported. The student should remember that the literature contains mention of a number of fungi, described or cited as species of *Sporotrichum* most of which are in reality synonymous with species of other genera such as *Beauveria* and *Acremonium* [e.g., *B. globulifera* (Speg.) and *A. tenuipes* Petch]. The genus *Rhinotrichum* contains at least one American species of entomogenous fungus (*R. depauperatum* Charles) parasitic on the avocado red mite. Members of the genus *Harmodendrum* are rarely pathogenic for healthy insects but are found occasionally on weakened or dead insects. In the United States, *Macrosporium* has been reported on certain Coleopetera, Lepidoptera, and Hemiptera. *Syngliocladium* has been found on wireworms.

The genus *Spicaria* contains numerous entomogenous species, one of which, *S. farinosa* (Fron) Vuill., was mentioned earlier in this chapter. It was with this species, as the type, that Petch (1934) proposed that the name *Isaria* be discarded in favor of *Spicaria, Isaria farinosa* (Disks.) Fron becoming *Spicaria farinosa* (Fron) Vuill. The generic name *Isaria* had been used generally by mycologists in connection with the conidial stages of *cordyceps*. About a dozen species of *Spicaria* have been reported as parasites of insects in North America. Among these, in addition to *S. farinosa*, are *S. heliothis* Charles on the corn earworm, *S. canadensis* Vuill. on larvae of the satin moth, and *S. rileyi* (Farl.) on a number of Lepidoptera and Coleoptera. The genus *Nomuraea* is generally considered as a synonym of *Spicaria*. Another genus whose species are imperfect stages of *cordyceps* is *Hymenostilbe*, of which at least six species have been found on North American insects.

## BASIDIOMYCETE INFECTIONS

The most remarkable association between insects and Basidiomycetes concerns those fungi belonging to the genus *Septobasidium*. Beneath the stromata of these fungi live certain scale insects, some of which are parasitized by the fungi. Because of the latter fact some authors consider the general relationship between the fungi and the insects to be one of parasitism. On the other hand, one of the leading students of the genus *Septobasidium*, Couch (1931), considers the relationship to be one of mutual symbiotism, in which the parasitization of some of the insects is for the good of the colony.

The number of known basidiomycete fungi truly parasitic on insects is comparatively small. Certain species in the genera *Helicobasidium, Corticium, Thelophora, Hymenochaete*, and *Daedalea* are now considered synonymous with members of the genus *Septobasidium*. Intermediate between the rusts and the genus *Septobasidium* has been erected the genus *Uredinella* of which the type species, *U. coccidiophaga* Couch, has been found on specimens of *Aspidiotus*. The imperfect genus *Hirsutella* was at one time thought to be a basidiomycete (Agaricales), but most of the species now contained in it are considered to be the imperfect stages of certain Ascomycetes, e.g., certain *Cordyceps*.

# 2

# ECTOTROPHIC ROOT-INFECTING FUNGI

This group of fungi includes nearly all the specialized root pathogens apart from the vascular wilt fungi. In the ectotrophic pathogens, infection of the root tissues is typically preceded by the superficial extension of mycelium over the root-surface; the interval between the advancing margin of epiphytic mycelium and the infection that follows it can vary from a millimetre or so with some pathogens of herbaceous plants to several metres with some fungi infecting roots of trees species. For *Rhizoctonia solani*, root infection by unspecialized fungal pathogens may also be preceded by superficial mycelial growth and attachment to the root surface. But because such unspecialized pathogens are usually limited by mature tissue-resistance to immature, senescent, wounded or unhealthy parts of the root system, this superficial spread of mycelium does not continue indefinitely; sooner or later, it is halted by host resistance. The specialized pathogens, on the other hand, are not thus halted by the development of effective defence mechanisms in older parts of the root system; provided always that soil conditions are favourable for ectotrophic infection, and essential proviso, they can continue to spread from the original foci of infection, perhaps on outlying parts of the root system, up to the root collar and stem base. The continuous, ectotrophic infection-habit of these specialized pathogens is thus no more and no less than the outward and visible sign of an infectivity that is adequate to overcome host resistance to infection throughout all parts of the root system.

There is an obvious superficial resemblance between the infection-habit of these specialized pathogens and that of the ectotrophic mycorrhizal fungi forming symbiotic associations with many species of forest tree; it was this resemblance that led me to propose the name 'ectotrophic' for these specialized pathogenic fungi as well. The use of this term 'ectotrophic' does indeed provide a convenient and unequivocal morphological description of the infection-habit of the pathogens but it cannot be taken to imply and relationship with the ectotrophic mycorrhizal fungi that is more profound than a similarity in morphology of the infection-habit. It can be suggested that the ectotrophic mycorrhizal fungi have evolved from the specialized root pathogens, according to the general postulates about the evolution of parasitism. No less plausible however, was Harley's (1948) suggestion that ectotrophic mycorrhizal fungi may have evolved from saprophytic root-surface fungi. Later in this chapter, we shall present much evidence for the conclusion that the ectotrophic infection-habit of the pathogens functions as an efficient mechanism for overcoming host resistance. Such a view is not appropriate for the ectotrophic infection-habit of the mycorrhizal fungi; in this apparently harmonious host-parasite relationship, there is no evidence that the fungus provokes any significant degree of host resistance, at least so far as the limited degree of infection that actually develops is concerned. Harley (1969) has described the morphogenesis and behaviour of various types of ectotrophic mycorrhizal association in critical detail. But a single example will suffice here to make my point. In his work on the ectotrophic mycorrhizal association of Scots pine, Robertson (1954) demonstrated that although infection of the short roots is truly ectotrophic (i.e. they are enveloped by the usual sheath of fungal mycelium), that of the long roots proceeds *internally*, as an extension of the Hartig net intercellularly in the cortex (Fig. 2.1).

From Robertson's evidence, it is clear that the ectotrophic infection-habit is not essential for successful extension of infection along the young long roots of Scots pine. We shall discuss later parallel situations with ectotrophic fungal pathogens, in which a decline of host resistance to a low level permits a purely internal extension of infection along roots of much impaired vigour. This suggests that the ectotrophic infection-habit is not impound upon ectotrophic mycorrhizal fungi, as it seems to be on ectotrophic pathogens, by the occurrence of a strong provoked resistance to infection. It thus seems most probable that the external fungal sheath of ectotrophic mycorrhizas has developed purely as an absorbing organ for the taking-up of nutrient salts from the soil. This

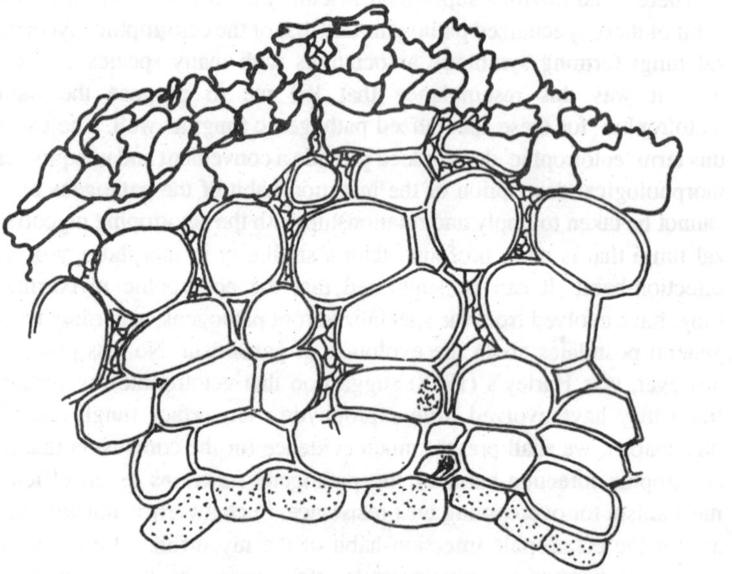


**Fig. 2.1.** Transverse section of cortex of mycorrhizal long root of *Pinus sylvestris* approx. 8 mm behind the growing apex. Note Hartig net of mycelium between cortical cells.

seems to have been a roundabout way for a plant pathologist to reach a conclusion that most mycorrhizal workers have probably always taken for granted.

## INFECTIVE MECHANISM OF THE ECTOTROPHIC INFECTION-HABIT

The essential features of the ectotrophic infection-habit amongst these specialized pathogens are well shown by *Ophiobolus graminis*, restricted to the Gramineae and causing the take-all disease of cultivated cereals. The fungus progresses at first epiphytically over the root system as a sparse network of dark-coloured runner hyphae, from which hyaline branches quickly penetrate and infect the underlying cortex and eventually the vascular cylinder of the roots (Fig. 2.2).

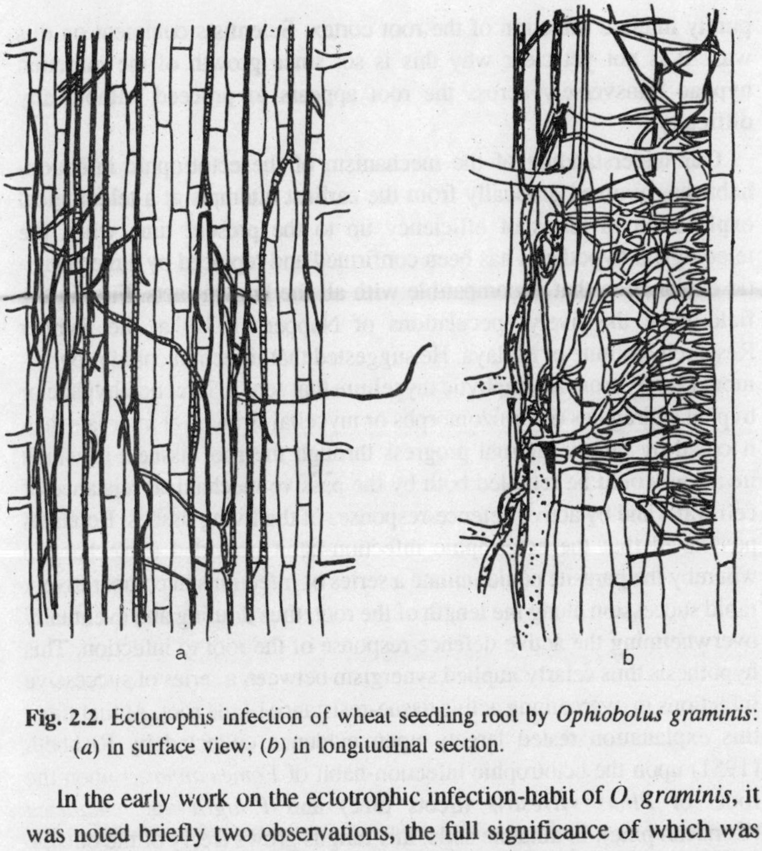

a b

**Fig. 2.2.** Ectotrophis infection of wheat seedling root by *Ophiobolus graminis*: (*a*) in surface view; (*b*) in longitudinal section.

In the early work on the ectotrophic infection-habit of *O. graminis*, it was noted briefly two observations, the full significance of which was not realized at that time. First, it was observed that as the fungus travelled down a seminal root from the point of inoculation below the germinating seed, it often abandoned the ectotrophic habit on the distal part of the root and the runner hyphae thereafter travelled intercellularly one or two cells deep within the cortex. Scientists attributed this to an impairment in resistance to infection of the distal part of the root, resulting from its virtual disconnection from the rest of the plant through destructive infection of the proximal part. The significance of this observation will become more apparent later, though the botanists did not appreciate its full implications at that time. Secondly, it was noted that if ectotrophic infection of a root was inhibited by sufficiently unfavourable conditions, such as too extreme a degree of soil acidity or too poor a degree of soil aeration, then *O. graminis* also failed to extend appreciably further by purely internal infection of the root cortex. Scientists comment on this was 'It is not yet clear why this is so, since growth of the infection hyphae transversely across the root appears to proceed without any difficulty'.

Our understanding of the mechanism of the ectotrophic infection-habit has evolved gradually from the earliest attempts at a teleological explanation in terms of efficiency up to the present time when the teleological hypothesis has been confirmed and replaced by a mechanistic explanation that is compatible with all the known facts. First in the field were the lively speculations of Napper (1934) at the Rubber Research Institute of Malaya. He suggested that infection could progress more rapidly from an epiphytic mycelium that sent in infection hyphae or hyphal aggregates (i.e. rhizomorphs or mycelial strands) at intervals than it could by purely internal progress through the root tissues, in which invasion would be retarded both by the passive mechanical reistance of cell walls and by active defence-responses of the living tissues. Botanists postulated that the ectotrophic infection-habit provided a mechanism whereby the parasite could initiate a series of infections in comparatively rapid succession along the length of the root, thus diluting and eventually overwhelming the active defence-response of the root to infection. This hypothesis thus celarly implied synergism between a series of successive infections in overcoming active tissue-resistance of the root. At that time, this explanation rested largely upon evidence collected by Rishbeth (1951) upon the ectotrophic infection-habit of *Fomes annosus* upon the roots of *Pinus sylvestris* (Scots pine) and *P. nigra* var. *calabrica* (Corsican pine). In alkaline soils, this fungus grows freely of the outside of pine roots, and the sheet-like mycelial aggregates amongst the bark scales can sometimes be found as much as half a metre ahead on infection in the wood cylinder. In acid soils (approx. pH 4.5), on the other hand, unfavourable soil conditions seem much to delay, and sometimes to suppress altogether, advance of the epiphytic mycelium, with the result that a slow progress of infection in the wood cylinder is sometimes ahead of epiphytic mycelium on the bark (Fig. 2.3).

Rishbeth's work with *F. annosus* thus laid the foundations for an understanding of the ectotrophic infection-habit; the critical experiment that has led to a satisfactory mechanistic explanation was carried out and published by one of Rishbeth's research associates. Wallis (1961). Wallis's experiments went very directly to the point at issue, i.e. whether the ectotrophic infection-habit is imposed upon the fungus by the active defence-response of the living root tissues. Wallis therefore made use of

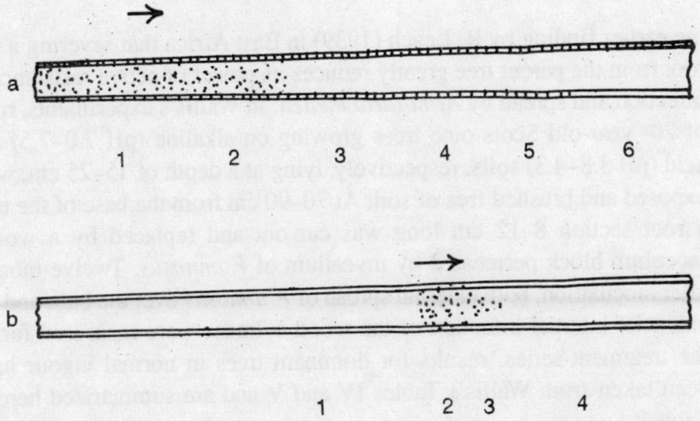


Fig. 2.3. Diagram representing two modes of infection of pine roots by *Fomes annosus.*

*(a) Infected root from alkaline soil*

| Region | Extent (cm) | Characteristics of region |
|---|---|---|
| 1 | >15 | Little resin exudation; wood dry; *F. annosus* hyphae abundant |
| 2 | 10 | Marked resin exudation; wood resinous; hyphae less abundant |
| 3 | 10 | Marked resin exudation; wood resinous; hyphae in bark only |
| 4 | 17 | Local resin exudation; wood not resinous; hyphae in bark only |
| 5 | 25 | Bark scales whitish, peeling; a few superficial hyphae present |
| 6 | — | Root normal, bark scales intact. |

*(b) Infected root from acid soil*

| Region | Extent (cm) | Characteristics of region |
|---|---|---|
| 1 | >15 | Bark split by resin; wood resinous; secondary fungi present only |
| 2 | 4 | Marked resin exudation; wood resinous; *F. annosus* hyphae in bark and wood |
| 3 | 2 | Slight resin exudation; wood resinous; hyphae in centre of wood only |
| 4 | — | Root normal |

Indicates region where *F. annosus* is present.

→ Indicates direction of invasion by *F. annosus.*

the earlier finding by R. Leach (1939) in East Africa that severing a tree root from the parent tree greatly reduces its power of active resistance to infection and spread by *Armillaria mellea*. In Wallis's experiments, roots of 20- year-old Scots pine trees growing on alkaline (pH 7.0–7.5) and acid (pH 3.8–4.3) soils, respectively, lying at a depth of 15–25 cm, were exposed and brushed free of soil. At 70–90 cm from the base of the tree, a root section 8–12 cm long was cut out and replaced by a woody inoculum block permeated by mycelium of *F. annosus*. Twelve months after inoculation, both external spread of *F. annosus* over the bark and the extent of internal infection of the wood cylinder were measured for all the treatment-series; results for dominant trees in normal vigour have been taken from Wallis's Tables IV and V and are summarized here in Table 2.1.

Wallis's results as summarized in Table 2.1 are so clear-cut as to need little comment. Behaviour of *F. annosus* on the proximal root portions, still connected with the parent tree and thus still in possession of full vigour and a full active defence-response to infection, is typical of that earlier observed by Rishbeth; ectotrophic advance of the fungus on the outside of the roots in alkaline soils is much ahead of that in acid soils, and it is only a matter of time before internal infection of the wood cylinder catches up with it. In the distal root portions, disconnection from the parent tree has evidently reduced the active defence-response to a level at which *F. annosus* has been able to spread more rapidly *within the wood cylinder* than it could do outside the root, even under the more favourable conditions for external spread, i.e. in the alkaline soil.

TABLE 2.1

**Extent of External Mycelial Spread and of Internal Infection of Wood Cylinder by Fomes annosus on Roots of Scots Pine (Measurements in cm)**

| | *Acid soil* | | *Alkaline soil* | |
|---|---|---|---|---|
| | *Outer bark scales* | *Wood* | *Outer bark scales* | *Wood* |
| Proximal root portions (connected) | 5 | 3 | 15 | 0.7 |
| Distal root portions (disconnected) | 15 | 48 | 56 | 75 |

From the results of Wallis's experiments, it has thus become clear that the ectotrophic infection-habit is imposed upon these specialized root pathogens by the active defence-response of root in possession of their full vigour. If no active defence-response is provoked, then internal infection can proceed without obvious hindrance, as we have already suggested in explanation of Robertson's (1954) observation on the internal extension of the Hartig net in the long roots of Scots pine (Fig. 2.1). The level of active defence-response of living root tissues varies to some extent in different host-parasite combinations. As we have noted earlier, it seems to be sufficient to prevent a purely internal extension of infection by *Ophiobolus graminis* within wheat roots. The same conclusion apparently holds for infection of roots of the rubber tree by *Fomes lignosus*; thus John (1958) demonstrated that infection of rubber roots by this fungus could be completely halted merely by removing the epiphytic rhizomorphs at weekly intervals whilst leaving the original inoculum undisturbed. In contrast to these two pathogens, *Fomes annosus* can sometimes progress slowly along a pine root by internal infection of the wood cylinder, even though such internal infections are quite often brought to a halt by the active defence-response of the living host-tissues; the nature of this defence-response of pine roots will shortly come up for discussion.

Following upon this recent work, it has now become possible to explain the mechanism of the ectotrophic habit on the assumption that the active defence-response of the host root is diluted and finally overcome by synergism between the effects of separate infections arising in succession from the externally travelling mycelium on the outside of the root. We can best illustrate this by an analogy with the epidemic development of the chocolate spot disease of field beans (*Vicia faba*), caused by species of *Botrytis* and chiefly by *B. fabae*. Wilson (1937) successfully elucidated the epidemiology of this disease by distinguishing between its *nonaggressive* and *aggressive* phases. The non-aggressive phase consists of scattered leaf-spots, or primary lesions, caused by single-spore infections; if the primary lesions are scattered thinly on the leaf, their further expansion is arrested by the host defence-response, and they remain static. But as Wilson demonstrated, one essential condition for development of the aggressive phase appears to be a degree of crowding of the primary lesions that exceeds a certain critical value; above this critical value, synergism between the effects of the individual infections operates in such a way as to overcome host resistance. The fungus is thus enabled to break out of the primary lesions and to initiate the aggressive phase of rapid spread through leaf and stem tissues, which continues unless and until interrupted by a spell of warm, dry weather. At the Cambridge

Botany School, we have confirmed and extended Wilson's findings in a yearly demonstration experiment for students set up by Mr. W.J. Bean, which has continued for nearly 20 years. In Fig. 2.4, we have provided a diagrammatic representation showing in parallel what we conceive to be the sequence of events following upon: (*a*) a single isolated infection; (*b*) a multiple series of infections either contemporaneous or shortly consecutive, first for the chocolate spot disease of beans and secondly for a typical, ectotrophic root-infecting fungal pathogen.

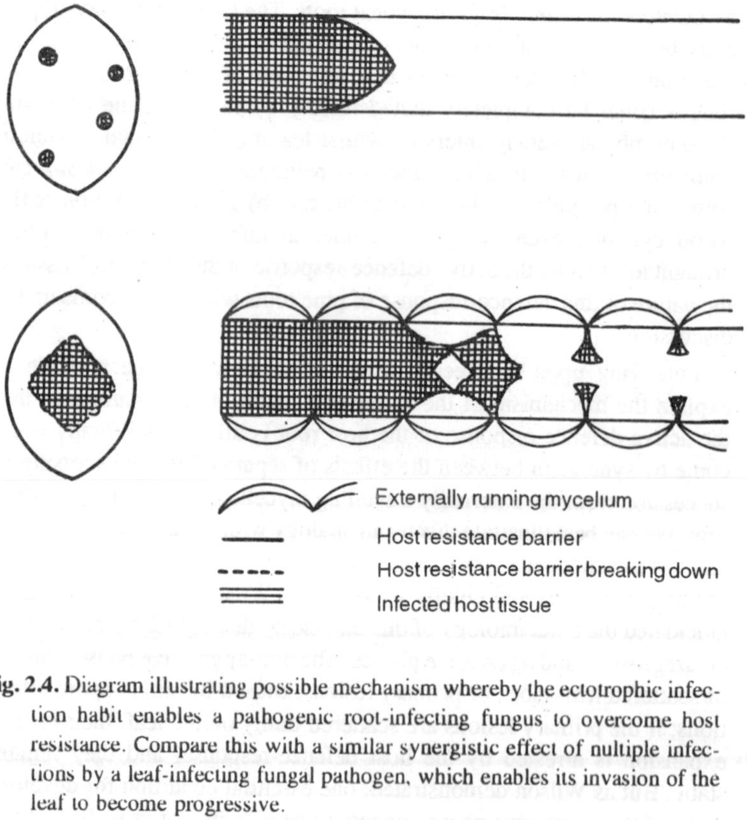


**Fig. 2.4.** Diagram illustrating possible mechanism whereby the ectotrophic infection habit enables a pathogenic root-infecting fungus to overcome host resistance. Compare this with a similar synergistic effect of nultiple infections by a leaf-infecting fungal pathogen, which enables its invasion of the leaf to become progressive.

## NON-ECTOTROPHIC INFECTIONS

Under this heading, it is appropriate to describe the behaviour of *Fomes annusus* on some species of conifer other than those of *Pinus*, as elucidated by Rishbeth (1951). In these other conifers, such as Douglas fir

(*Pseudotsuga taxifolia*), larch (*Larix decidua*) and spruce (*Picea abies*), resistance to infection of the inner wood tissues is lower than that in species of pine, and so infection follows a very different course. In these conifers, as in pines, exposure of very young trees to heavy or multiple infections can result in outright killing; in older trees, however, infection leads to development of but rot, which can cause much heavier losses of marketable timber than does the more dramatic tree-killing as manifested in pine plantations. In trees, some 20–30 years old, of Douglas fir, larch or spruce, the following infection sequence has been described by Rishbeth. The youngest roots are generally the first to become infected, usually by contact with infected stump roots resulting from basidiospore infection by *F. annosus* of the tree stump surfaces, exposed by tree felling during a plantation thinning. In these small roots, there is little resin response by the root to the developing infection, and *F. annosus* grows rapidly in both bark and wood until it reaches an older part of the root of some 2 cm in diameter. Thereafter, resistance of the central wood is lower than that of the outer sapwood, and so the fungus advances through an inner cylinder of wood, leaving the outer sapwood uninfected and still able to produce branch roots (Fig. 2.5).

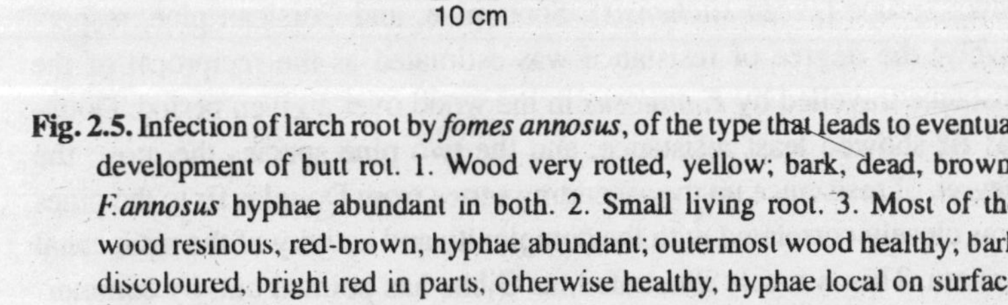


**Fig. 2.5.** Infection of larch root by *fomes annosus*, of the type that leads to eventual development of butt rot. 1. Wood very rotted, yellow; bark, dead, brown; *F. annosus* hyphae abundant in both. 2. Small living root. 3. Most of the wood resinous, red-brown, hyphae abundant; outermost wood healthy; bark discoloured, bright red in parts, otherwise healthy, hyphae local on surface only. 4. As last, but smaller region of central wood with hyphae. 5. Whole root healthy.

As *F. annosus* approaches the root collar, or butt of the tree as it is sometimes called, the boundary between the central cylinder of infected wood and the outer uninfected wood is often marked by a conspicuous layer of resin (Fig. 2.6).

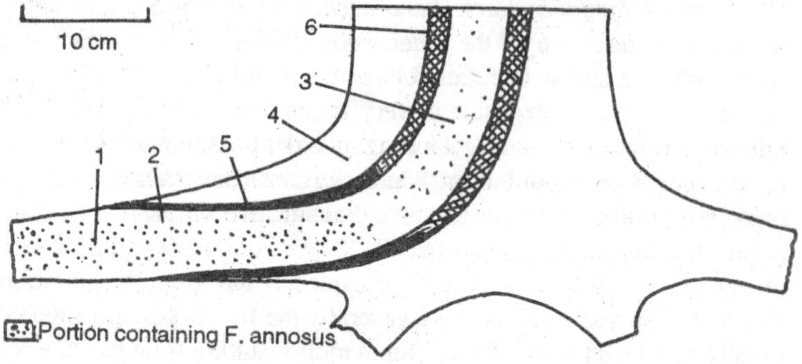


**Fig. 2.6.** Diagrammatic section of lower butt of Douglas fir, showing incipient development of butt rot, with *Fomes annosus* entering from a lateral root. 1. Wood extensively rotted, yellow. 2. Wood, partially rotted, yellow-brown. 3. Wood with incipient rot, red-brown. 4. Wood healthy, white. 5. Resin layer. 6. Wood resinous, deeply stained.

It is thus apparent that in species of conifer susceptible to the butt-rot type of infection, internal spread of *F. annosus* is made possible by the fact that the resistance of the inner cylinder of wood is sufficiently lower than that of the outer sapwood in these species, and is also lower than that of any part of the wood cylinder in pine species. The reasons for these differences have been elucidated in an investigation of outstanding interest by Gibbs (1968), working as one of Rishbeth's research associates. By direct inoculation of *F. annosus* into the wood cylinder of the roots, Gibbs was able to estimate the resistance of the wood in Douglas fir, Sitka spruce (*Picea sitchensis*), Scots pine, and Corsican pine, respectively; the degree of resistance was estimated as the reciprocal of the distance travelled by *F. annosus* in the wood over a given period. Douglas fir showed least resistance, and the two pine species the most; the degree of resistance up the ascending series from Douglas fir to the pines was closely correlated with the complexity and activity of the resin canal system. This is nicely illustrated, as Gibbs has pointed out, by commercial use of these various genera of conifers for commercial production of resin; various species of *Pinus* are widely employed for resin tapping, those of *Picea* much more rarely, and those of *Larix* and *Pseudostuga* scarcely at all. Gibbs further noted from the results of his inoculation experiment that in spruce and Douglas fir, *F. annosus* failed to infect the outer sapwood and was restricted to the inner wood, in which a high proportion of the resin ducts become atrophied. In other experiments, with Scots and Corsican pine, Gibbs recorded daily resin yields by tapping resin yield, tree vigour and degree of resistance to *F. annosus* all varied in parallel under the conditions obtaining in these trials.

An analysis of the toxicity of resin (in this chapter designated as 'oleoresin') to *F. annosus* and four timber bluestain fungi (*Ceratocystis* spp.) has been made by Cobb, Krstic, Zavarin & Barber (1968). They found that both crude oleoresin and nine components of its volatile turpentine fraction, purified by fractional distillation, reduced mycelial growth of *F. annosus* and of all four bluestain fungi by various degrees when the fungi were exposed to a vapour-saturated atmosphere; growth reductions were still greater when either oleoresin or one of its volatile components were incorporated in the growth medium. These nine turpentine components were identified; there were two monocyclic, four bicyclic and one open-chain terpene, and two alkanes. The most fungistatic of these components was the alkane *n*-heptane, which in a vapour-saturated atmosphere completely inhibited mycelial growth of both *F. annosus* and *C. pilulifera*.

## MYCELIAL AGGREGATION INTO SHEETS, STRANDS AND RHIZOMORPHS

Ectotrophic pathogens that infect woody perennials, i.e. bushes and trees, typically show aggregation of the epiphytic mycelium into mycelial sheets or strands, or into rhizomorphs. They thus contrast with fungi infecting herbaceous species, such as *Ophiobolus graminis*, in which the epiphytic mycelium is not organized into hyphal aggregates but consists of separated individual hyphae. This distinction is not affected by the fact that *Armillaria mellea*, a typical ectotrophic parasite of tree root systems, will infect potato tubers if opportunity offers; were *A. mellea* limited to potato plants and other herbaceous species, it is unlikely that it would have evolved so powerful an offensive armament as the fungal rhizomorph. Evidence of three kinds suggests that these various forms of hyphal aggregation have evolved in response to the need for an inoculum potential adequate to overcome host resistance to infection by the roots of woody perennials, and particularly by those of the larger arborescent species. First, mycelial aggregates as organs of infection are not found in root-infecting fungi restricted to herbaceous hosts. Secondly, many experiments have now shown that there is a critical level of inoculum potential necessary for initiation of a progressive infection;

the older, larger and more highly organized host roots demand a higher level of inoculum potential for establishment of a progressive infection than do younger, smaller and simpler roots. Thirdly, there is no evidence for the supposition that the older, cork-covered parts of tree root systems can be successfully infected either by spores or by individual hyphae in unorganized mycelium, though there is much evidence for the contrary view that infection can be accomplished only by mycelial aggregates growing out from a food base generating adequate inoculum potential.

## MORPHOGENESIS AND BEHAVIOUR OF MYCELIAL STRANDS AND RHIZOMORPHS

Mature mycelial strands are superficially difficult to distinguish from rhizomorphs; both types of organ may reach a diameter varying from one to several millimetres, and a mature strand may have a surface as smooth as that of a rhizomorph. These two types of fungal organ, however, develop in very different ways and whether a particular organ is a strand or a rhizomorph can be determined only by a careful study of its morphogenesis. A mycelial strand is built up gradually, sometimes by the chance meeting of individual hyphae or minor strands; more commonly, however, strands develop acropetally by the ensheathing growth of branches of the parent leading hypha or haphae. At least two types of strand morphogenesis have been described, and further variants on the general theme are likely to be discovered. Rhizomorphs, on the other hand, are morphogenetically similar to the roots of a higher plant; they grow from an apical meristem, though this meristem is of a type quite different from that found in any higher plant.

In consequence of this difference in morphogenesis, mycelial strands and rhizomorphs also differ in the way that their inoculum potential develops. The diameter of a rhizomorph attains its maximum value just behind the growing apex, and so the apical region possesses, to the maximum possible degree, 'instant inoculum potential'. A mycelial strand, in contrast, makes its first contact with a host surface with the arrival of its leading hypha (or associated leading hyphae); thereafter, both strand diameter and inoculum potential increase acropetally up the still developing strand, as further constituent hyphae grow up towards the strand apex from the rear.

**Mycelial Strands**

The morphogenesis of the mycelial strands produced by *Helicobasidium purpureum*, causing the violet root rot disease not only of sugar beet but also of various woody perennial species, has been analysed by Valder (1958). Strands develop from a food base lying in or on the soil when environmental conditions are suitable for mycelial growth. The food base employed by Valder in this study was a disc of nutrient agar colonized by *H. purpureum*, but under natural conditions both infected roots and sclerotia provide food bases for strand development. From such a food base, there first develops a sparse growth of robust, coloured hyphae over and through the soil; further hyphae growing out from the food base sooner or later encounter one of these leading hyphae and then follow it in associated growth. By continued accretion of these 'following' hyphae and also of small strands, each of the original leading hyphae may develop into a main strand, the base of which comes to resemble a river delta. Owing to this mode of morphogenesis, these strands are widest at the base and taper off towards the apex. Anastomoses occur between strands to give a network, especially near the food base. Branching of strands can occur, due to the branching of the original leading hypha, with the branch attracting some of the follower hyphae away from their parent hypha. Valder further observed that coherence of the strands of *H. purpureum* was brought about in three ways: (1) by an interweaving growth of the associated main hyphae; (2) by the binding action of short branches of limited growth, sometimes hooked at the apex; (3) by anastomoses between hyphae (Fig. 2.7).

A second and commoner type of strand morphogenesis is exemplified by that of *Phymatotrichum omnivorum* (Shear) Duggar, which causes Texas root rot of cotton and other dicotylendonous species, including a number of woody perennials. Rogers & Watkins (1938) have described how the strands of this fungus develop around single, exceptionally wide hyphae, which become ensheathed by their own branch hyphae. By subsequent adjustment of position and septation in the ensheathing branch hyphae, a compact cortex of several layers in thickness is eventually formed around the single large hypha at the centre of each strand (Fig. 2.8).

A type of strand morphogenesis very similar to that in *P. omnivorum* has been described for two well known saprophytes: *Merulius lacrymans* Fr., causing dry rot of house timber, and *Agaricus bisporus* (J. Lange) Pilát, the cultivated mushroom.

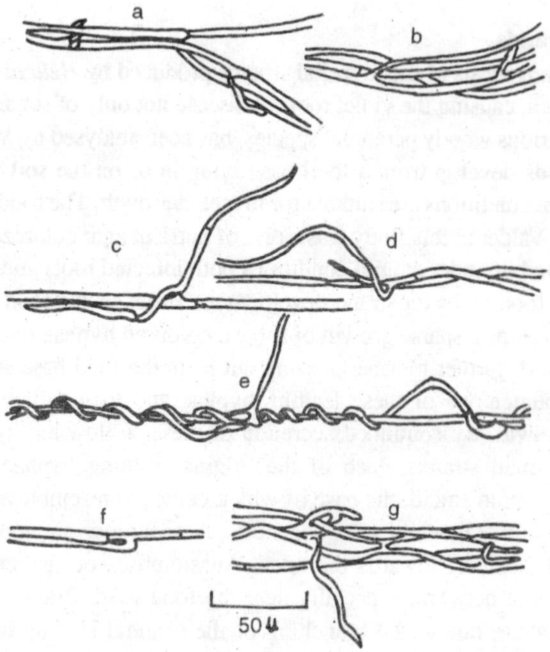


**Fig. 2.7.** Morphogenesis of mycelial strands in *Helicobasidium purpureum*. The General direction of hyphal growth is from left to right. Note interweaving growth of main hyphae, binding action of short side-branches of limited growth (*a, b, g*), and the locking action of hyphal anastomosis (*g*).

Mycelial strands are typically the form in which a strand-producing fungus migrates outwards from a food base through the mineral matrix of the soil, which approximates to a non-nutrient medium, and at first epiphytically over root surfaces. When growing on a nutrient medium, such fungi either do not produce distinguishable strands, except amongst the aerial mycelium out of contact with the substrate, or they produce them only during the later stages of substrate colonization. In earlier accounts of strand morphogenesis, Botanists had postulated an attraction between hyphae of the same mycelium or species as responsible for strand morphogenesis, and had suggested that this attraction might be similar to, or even identical with, that responsible for anastomoses between hyphae. Absence of strands, or delayed production, on nutrient media had then to be explained by a repulsion between hyphae in such a situation that was strong enough to override the postulated attraction. Causes of repulsion were not far to seek; on a nutrient medium every hypha acts as a diffusion sink for nutrients and as a source of growth products, some of which are likely to become fungistatic above a certain concentration. The consequences of these two processes were thought to cause hyphae to keep as far away as possible from one another on a nutrient medium, and thus to favour a dispersed type of growth.

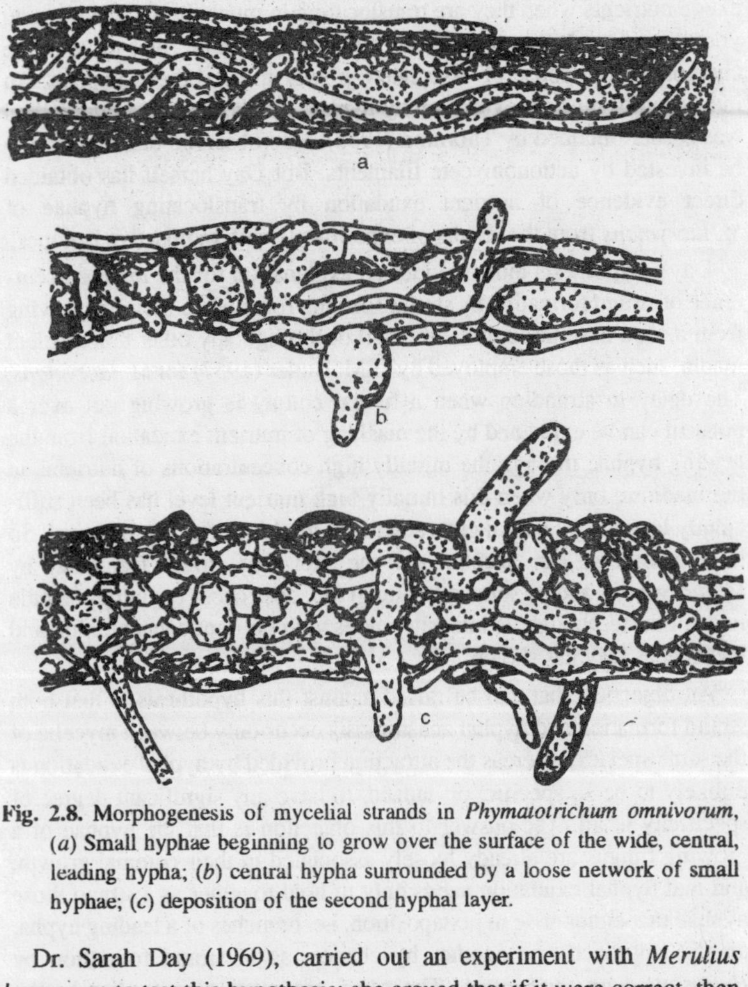


**Fig. 2.8.** Morphogenesis of mycelial strands in *Phymatotrichum omnivorum*, (*a*) Small hyphae beginning to grow over the surface of the wide, central, leading hypha; (*b*) central hypha surrounded by a loose network of small hyphae; (*c*) deposition of the second hyphal layer.

Dr. Sarah Day (1969), carried out an experiment with *Merulius lacrymans* to test this hypothesis; she argued that if it were correct, then *M. lacrymans* should form strands in a liquid nutrient medium that was renewed sufficiently rapidly by a drip feed device to prevent diffusion shells, either of nutrients or of fungal growth products, forming around individual hyphae. The results of this experiment were the reverse of those expected; mycelial strands formed in the still but not in the replacement cultures. From a consideration of these results and of many observations by earlier workers, Day was able to formulate a new hypothesis that seems to account satisfactorily for all the known facts. This hypothesis postulates that the leading hyphae, around which strands are formed, exude nutrients when they are translocating in much the same way as do young root and that it is this nutrient exudation that causes the chemotropic response of associated growth in the branch hyphae. An indication of nutrient exudation by hyphae of *Rhizoctonia solani* in soil was earlier obtained by Thornton (1953), who observed these hyphae to be invested by actionomycete filaments. But Day herself has obtained direct evidence of nutrient exudation by translocating hyphae of *M. lacrymans* from the results of experiments with radioactive isotopes.

Day's hypothesis thus provides an explanation for the regular occurrence of strand formation by strand-forming fungi when they are growing from a food base through natural soil or through any other non-nutrient media, such as those employed by G.M. Butler (1957) for *M. lacrymans*. The delay in stranding when a fungal colony is growing out over a nutrient can be explained by the masking of nutrient exudation from the leading hyphae through the initially high concentrations of nutrients in the medium; only when this initially high nutrient level has been sufficiently lowered by fungal growth will the masking effect be removed. So this hypothesis can account for the fact that strand formation by *M. lacrymans*, and by some other fungi too, does occur on nutrient media in due course, though its intensity varies with nutrient concentration and balance, as Day has shown.

An objection that can be raised against this hypothesis is that both strand formation and hyphal anastomosis occur only between mycelia of the same species, whereas the attraction provided by hyphal exudation is unlikely to be so specific, or, indeed, to have any significant degree of specificity at all. The answer to this objection is that the hyphae of a growing fungus are already loosely associated in their colonial growth, and that hyphal exudation needs only to hold together as a strand those hyphae that either arise in juxtaposition, i.e. branches of a leading hypha, or else approach one another by chance, as in strand formation by *Helicobasidium purpureum*. The same argument also applies to the mechanism of anastomosis; a nutrient stimulus from hyphal exudation will serve to prolong a hyphal association brought about by mere chance in the first place. Indeed, hyphal association represents the first stage only in the development of anstomosis; as Flentje (1969) has described for *Rhizoctonia solani*, various causes of failure are spaced in the time sequence from initial juxtaposition of hyphal growth up to the ultimate consummation of anastomosis between compatible hyphae.

**Rhizomorphs of Armillaria Mellea**

The best known rhizomorph is that produced by *Armillaria mellea*. If we assume the mean diameter of the straight running, unbranched, constituent hyphae to be about 10 μ, then we can calculate that a rhizomorph of 1 mm diameter will be made up of several thousand hyphae, even after allowing generously for inter-hyphal space and for the central open channel running down the mature part of the rhizomorph. Growth of the rhizomorph, like that of a root, occurs from an apical meristem, and this implies a mechanism for coordinating apical growth of the constituent hyphae. When a fungal colony is growing on a nutrient agar, the rhizomorphs produce branches and the density of branching increases with the concentration of nutrients in the medium up to an optimum value. On such a nutrient medium, every rhizomorph eventually produces a mantle of fringing mycelium, which further enhances its resemblance to a root with its mantle of root hairs. Fig. 2.9, after R. Hartig, shows the apical region of a rhizomorph of *a. mellea.*

Much has been learnt about the behaviour of rhizomorphs of *A. mellea* through simple experiments on the nutrient agar plate. When a 4 mm agar disc taken from the growing margin of a colony of *A. mellea* is laid at the centre of a fresh plate of nutrient agar, an outwards growth of unorganized mycelium can be observed within 24 h at 25°C; after 7 days, a circle of rhizomorph initials can be seen around the margin of the original inoculum disc. In an experiment, the young rhizomorphs, which started growth later than the colony of unorganized mycelium, grew some five times as quickly as the mycelial colony, and hence rapidly outgrew it. From the older region of the rhizomorphs fringing mycelium grew out, so that a mature rhizomorphic colony on an agar plate was deeply lobed. These observations were confirmed and amplified by Snider (1959), working with a number of single-basidiospore isolates of *A. mellea*. Undoubtedly the most interesting fact to emerge from these experiments was the wide difference in growth rate between that of hypahe in unorganized mycelium and that of the rhizomorphs; in a subsequent study by Rishbeth (1968), rhizomorphs of one isolate grew some thirteen times as quickly as the margin of the mycelial colony. A possible explanation for this difference is that the constituent hyphae

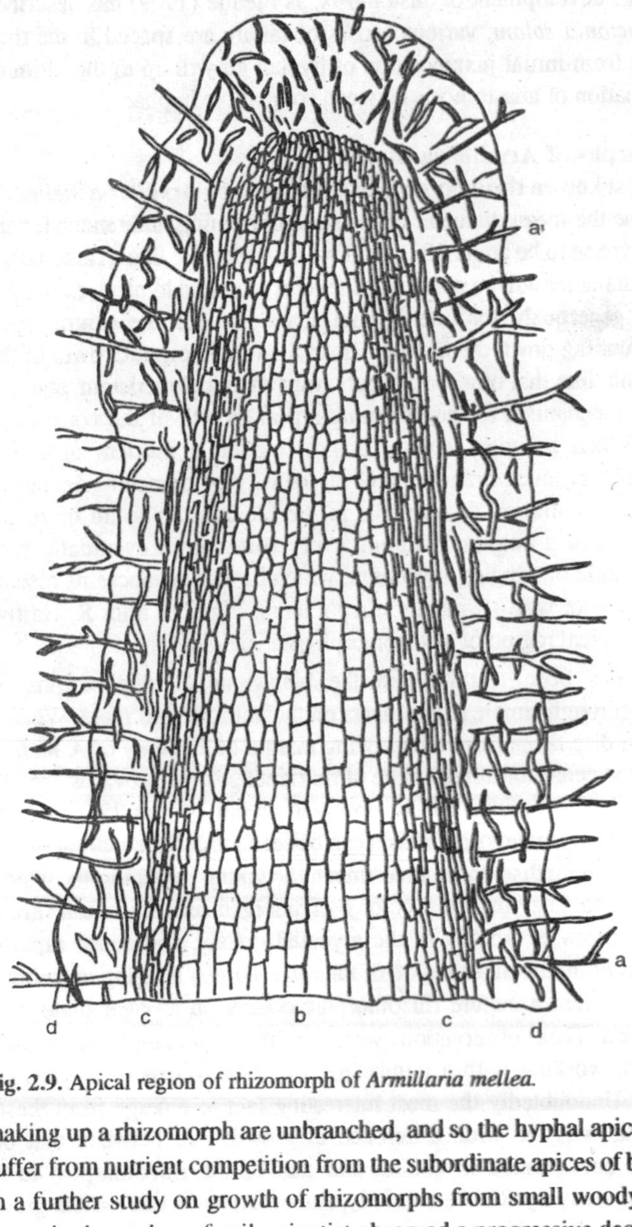


Fig. 2.9. Apical region of rhizomorph of *Armillaria mellea.*

making up a rhizomorph are unbranched, and so the hyphal apices do not suffer from nutrient competition from the subordinate apices of branches. In a further study on growth of rhizomorphs from small woody inocula through glass tubes of soil, scientist observed a progressive deceleration with time in the growth rate of rhizomorphs, and obtained evidence that this was due to increasing competition for nutrients from the subrodinate apices of rhizomoprhs branches. But whatever the reason for this very high growth rate of rhizomorphs, as compared with that of individual hypahe in unorganized mycelium, it must have one important consequence for the mechanism of ectotrophic infection. If this interpretation of this mechanism, as illustrated in Fig. 2.4, be accepted then the faster the rate of epiphytic spread of the rhizomorphs, the more effective will be the ectotrophic infection-habit in overcoming active host resistance.

Another outstanding difference in morphogenesis of mycelial strands and rhizomorphs has also become apparent; whereas strand-building is delayed by the presence in the growth medium of nutrients above a low concentration, the production of rhizomorph initials increases with glucose concentration at least up to the level of 2%. In an experiments on production of rhizomorphs on nutrient agar, vitamins were provided by addition of thiamin and also by a supplement of malt extract and/or peptone, to guard against any possible vitamin deficiency. Employing a synthetic growth medium with 1 mg thiamin/l as the only vitamin supplement, Weinhold & Garraway (1966) found that rhizomorphs were not produced either in the liquid medium or on the same solidified with agar unless the medium was further supplemented by ethanol, at the rate of 500 ppm. As the medium was provided with 6 g. glucose/l as a carbon source, the ethanol supplement was evidently acting as a rhizomorph-inducing factor rather than as carbon source, even though ethanol by itself at the rate of 4.6 g/l provided a satisfactory carbon source for both mycelial growth and rhizomorph production. Mycelial colonies grew well enough in the absence of ethanol, though growth was usually improved by its addition to the medium. The isolation of *A. mellae* grew well enough in a synthetic liquid medium with thiamin as the only vitamin supplement, but rhizomorphs were virtually absent. Scientist observations with this single isolate were confirmed by Azevedo (1963) with two isolates from England and two from the Azores, respectively. For some isolates, at least, of *A. mellea*, ethanol and other low-molecular weight alcohols thus seem to act specifically as a rhizomorph-inducing factor, though their experience suggests that they can be replaced by some unknown factor or factors present in malt extract and in peptone. The results of further experiments by Garraway & Weinhold (1968) showed that for maximum mycelial growth and rhizomorph production, their isolates of *A. mellea* required ethanol for the first 15 days of colony growth; thereafter, lack of ethanol effected neither rhizomorph production nor total growth.

Production of rhizomorphs is not invariably associated with activity of *A. mellea* as a root-infecting pathogen. Their complete absence from areas of root disease in cacao plantations in Ghana was noted by Dade (1927) and their rarity in Malawi by R. Leach (1939). They are abundantly produced in the forests of the Kenya Highlands but are seldom found in the forests of Rhodesia surveyed by Swift (1968). Where rhizomorphs are thus absent, *A. mellea* spreads from infected to healthy roots by direct contact, and thereafter by mycelial sheets in the bark and over the plane of the cambium; the mechanism of this spread as an alternative to ectotrophic infection by initially epiphytic rhizomorphs clearly calls for further investigation. In an endeavour to elucidate this variability in the production of rhizomorphs by *A. mellea* in different areas, Gibson (1961) studied 140 isolates of *A. mellea* from various parts of the world and found that they varied widely in the regularity, abundance and vigour of rhizomorph production under optimum conditions of pure culture on nutrient agar. A total of 116 isolates was tested for their temperature optima over the three ranges of 22–4°C, 25–7°C and 27–9°C, respectively. Both for mycelial growth and for initiation and growth of rhizomorphs, most isolates had their optima within the range 25–7°C; the frequency distribution of temperature optima did not differ significantly between isolates from temperate and from tropical regions, respectively. These results thus seemed to give little support to the idea that production of rhizomorphs under natural conditions was limited by soil temperature, i.e. that too high a temperature prevented their initiation at low altitudes in the tropics, but not at higher elevations in the tropics nor in temperate climates.

Fortunately, however, this proposition has recently been reinvestigated by Rishbeth (1968), who has thereby brought to light a critical difference between temperature optima for initiation and growth of rhizomorphs on nutrient agar and from woody inocula through soil, respectively. Growth rate both of purely mycelial colonies and of rhizomorphs on 3% malt extract agar was found to be maximal at a temperature of approx. 25°C; growth rate of rhizomorphs from woody inoculum segments through soil tubes was maximal at approx. 22°C (Fig. 2.10).

The optimum temperature for growth rate of rhizomorphs from a food base through soil is thus lower, by some 3°C, than the optimum for growth of rhizomorphs on malt-extract agar. Botanist found that growth rate of rhizomorphs from woody inoculum segments through soil tubes is nutrient-dependent. Growth is faster from large inocula than from small

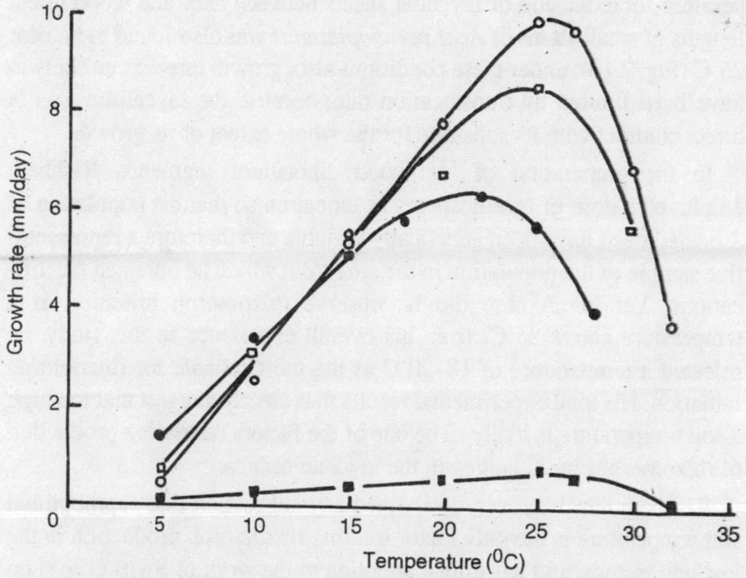


**Fig. 2.10.** The growth rates of mycelium and rhizomorphs of *A.mellea* in relation to temperature. O, Rhizomorphs in malt agar; •, rhizomorphs in soil tubes; □, mycelial sheets in wood lengths; ●, mycelium in malt agar.

ones, and weekly growth increments of rhizomorphs invariably decrease with time from the beginning of growth; They suggested that this latter effect was due partly to a decrease in rate of mobilization of nutrients from the woody food base, and partly to an increasing intensity of competition for nutrients between the main apices of the rhizomorphs and the growing number of their subordinate branch apices. Rishbeth has used this evidence in support of his suggestion that at the optimum temperature of approx. 22°C, rhizomorph growth rate is in balance with translocation from the food base; at higher temperatures, rate of translocation is insufficient for the higher rate of metabolism in the rhizomorph apices. This suggestion was supported by the effects of transferring soil tubes from a lower to a higher temperature and vice versa. It also explains Rishbeth's observation that the optimum temperature for growth of rhizomorphs on malt-extract agar was 25°C and sometimes higher; under these growth conditions, rhizomorph growth rate would be unlikely to be nutrient-limited, because uptake of nutrients can take place along the whole length of the growing rhizomorph. The optimum temperature for extension of mycelial sheets between bark and wood in cut lengths of small stems of *Acer pseudoplatanus* was also found to be near 25°C (Fig. 2.10); under these conditions also, growth rate was unlikely to have been limited by translocation rate, because the mycelium was in direct contract with its substrate for the whole extent of its growth.

In the preparation of his woody inoculum segments, Rishbeth employed a dose of basidiospores as inoculum so that his population of *A. melea* must have been genetically variable and therefore a representative sample of the population in the area from which he obtained fructifications. Yet in no case did he observe rhizomorph initiation at a temperature above 25°C; from his overall experience in this study, he selected a temperature of 18–20°C as the most reliable for rhizomorph initiation. His total experimental results thus clearly suggest that too high a soil temperature is likely to be one of the factors restricting production of rhizomorphs by *A. mellea* in the lowland tropics.

Rishbeth has, however, disclaimed any suggestion that supraoptimal soil temperature is the only factor limiting rhizomorph production in the lowland tropics, and has drawn attention to the work of Swift (1968) on the existence of a toxic factor in some Rhodesian forest soils. All the isolates of *A. mellea* employed by Swift produced abundant rhizomorphs on nutrient agar, so that an intrinsic inability to produce rhizomorphs could be eliminated as a possible factor in their behaviour. Moreover, although no rhizomorphs were produced from woody inoculum segments in growth tubes filled with a typical Rhodesian forest soil, they were initiated and grew well in the same soil after it had been autoclaved. Soil fumigation with propylene oxide did not remove the rhizomorph-inhibiting factor, so Swift was able to conclude that it was persistent in the soil in the absence of living micro-organisms, at least for a time, through it was not thermostable because it was destroyed by autoclaving the soil. The persistence of this factor was also demonstrated in another way, by Seitz-filtration of a 1 : 1 soil-water extract and addition of the sterile filtrate to columns of autoclaved soil in the growth tubes; this sterile filtrate caused total inhibition or rhizomorph growth.

Swift also considered another possibility, arising from an observation that rhizomorphs failed to make any appreciable growth into the short column of moist sand with which scientists covered inoculum segments, standing above soil columns in the growth tubes, in some of their experiments. They thought that perhaps sand failed to provide some factor present in soil that stimulated outwards growth of rhizomorphs from the inoculum segments; this possibility was later eliminated by Rishbeth's (1968) finding that rhizomorphs grew well enough from the inoculum segments when these were buried in 1 lb glass jars of sand. A satisfactory resolution of this discrepancy has now been provided by Griffin (1969), who has suggested that the earlier observations can be explained by a drying-out of the short sand columns covering the inoculum segments at the top of the growth tubes. In this supposition Griffin (1969) has been entirely correct; It could indeed be confirmed that the short sand columns did dry out fairly quickly, in spite of the fact that the growth tubes were capped by moisture-proof, lacquered cellophane secured with a rubber band. The correct explanation did not occur to them because the drying sand still acted as an effective mulch to prevent drying out of the inoculum segments, and growth of rhizomorphs *downwards* into the soil columns was very satisfactory. But failure of rhizomorph initials to continue growth into the drying sand columns can now be explained by Griffin's discovery, in collaboration with A.M. Smith, that rhizomorphs will continue to grow only for so long as the apices are covered by an unbroken film of water. They have interpreted this requirement in terms of the reduced rate of melanin formation under the lower oxygen concentration around the rhizomorph apex when it is covered by a water film; their explanation is compatible with the high values of the Michaelis constants for the enzymes laccase and tyrosinase. They postulate that the melanin sheath is mycostatic, protecting the mature part of the rhizomorph but inhibiting elongation of the apex if formed precociously.

## SOME EFFECTS OF SOIL CONDITIONS ON ECTOTROPHIC SPREAD

Three ectotrophic root-infecting fungi are known to spread more quickly over the root systems of their hosts in alkaline soils than in neutral or acid ones. No explanation seems to have been advanced for this effect on the spread of *Phymatotrichum omnivorum*, but attempts have been made to explain a similar effect on the spread of *Ophiobolus graminis* and *Fomes annosus*, respectively.

During an investigation of the effect of soil conditions on rate of ectotrophic spread by *Ophiobolus graminis* along the roots of wheat seedlings, scientists found that rate of spread was increased: (1) by improvement in soil aeration; (2) by increase in pH value of the soil up to approx. pH 8. In a hypothesis attempting to unify these two sets of observations, they suggested that rate of advance of *O. graminis* was inversely proportional to the concentration of carbon dioxide at the surface of the respiring root; thus at pH 8.3% undissociated carbon dioxide is in equilibrium with 97% as bicarbonate ion, whereas at pH 5.5, 90% free carbon dioxide is in equilibrium with 10% as bicarbonate. In further experiments to test this hypothesis, they demonstrated that growth of *O. graminis* along the roots was greatly accelerated by forced aeration of the soil and that under these conditions soil pH value no longer affected the rate of fungal advance. This simple explanation has recently been challenged by the theoretical considerations and experimental results published by Greenwood & Nye (1968). They have pointed out that as diffusion through the gas phase is extremely rapid, oxygen and carbon dioxide partial pressures in any soil gas space that is directly connceted with the atmosphere above the soil will be much the same as in air. In the soil-water phase, carbon dioxide is thirty times as soluble as oxygen and so will diffuse thirty times as fast. From this it could be concluded that a deficiency of oxygen at the root surface, due to both root and microbial respiration, is likely to limit the growth rate of *O. graminis*, and of similar ectotrophic root-infecting fungi, *before* the partial pressure of carbon dioxide has increased to a growth-limiting value. Nevertheless, cónvincing as these arguments by Greenwood & Nye may seem, it is perhaps yet premature to come to any definite conclusion; their arguments challenge not only my own hypothesis but also a substantial volume of other data hitherto accepted as evidence for a growth-limiting effect of carbon dioxide on the activities of various soil micro-organisms.

A different explanation was put forward by Rishbeth (1950) for the more rapid advance of *Fomes annosus* over the bark of pine roots in alkaline as compared with acid soils. He was able to reproduce this effect on 5 cm lengths cut from small pine roots growing in acid and alkaline soils, respectively, washed free of soil, and then incubated, after inoculation with *F. annosus* at one end, in a moist atmosphere. The design of this experiment eliminated any differential effect of carbon dioxide at the root surface, such as scientist had postulated for *O. graminis*. It must also largely have eliminated any differential effect of active host resistance due to soil type, because as we have noted earlier, Wallis (1961) demonstrated that root-severing destroyed the effective defence-response of roots in either type of soil. The much retarded rate of advance by *F. annosus* over the root bark of roots taken from an acid soil in this experiments was ascribed by Rishbeth to antagonistic effects by root-surface fungi of acid soils, and particularly to those exerted by *Trichoderma viride* Pers. ex. Fe. Gibbs (1967) has examined Rishbeth's suggestion of *T. viride* as an important antagonist of *F. annosus* on pine root bark in acid soils in the light of Rifai's (1969) subdivision of *T. viride*, as recognized by Bisby (1939), into nine species. Even within each of Rifai's narrowly defined species of *Trichoderma*, both Gibbs (1967) and Mughogho (1968) have found a wide variation amongst isolates in the degree of antagonistic effects that they exert against other fungi growing in close association with them. Gibbs was unable to find any clear evidence of a significant correlation between frequency of antagonistic isolates of *Trichoderma* and soil acidity. In place of this explanation, he has drawn attention to much published evidence in favour of the generalization that most antibiotics are inactivated much more quickly in neutral and alkaline soils than they are in acid soils.

On the light-textured, sandy soils of the Breckland area of East Anglia particularly studied by Rishbeth (1951), various other soil factors, and particularly organic content, can be closely correlated with soil pH. Thus the acid podsols are covered by a well marked litter layer, and so the moisture retaining and supplying power of these soils in dry seasons is higher than that of the alkaline sandy soils. Rishbeth adduced both observational and experimental evidence for the view that the more favourable moisture regime in the acid podsols enhanced the resistance of pine trees to *F. annosus*, especially in dry seasons. This opinion has been upheld by recent evidence produced by Gibbs (1968), showing that the volume of resin production following injury could be directly correlated with soil moisture availability, and was greater for pine trees on acid podsols than for those on the alkaline sandy soils. In the USA, Towers & Stambaugh (1968) found that experimentally induced moisture stress increased the susceptibility of lobolly pine (*Pinus taeda*) to infection by *F. annosus*.

# 3

# SAPROPHYTIC ROOT-INFECTING FUNGI

This chapter and the two following will be concerned with survival of pathogenic root-infecting fungi in the soil apart from living host plants. There are three possible modes of such survival: (1) by competitive saprophytic colonization of dead organic substrates, most commonly consisting of a corpus of dead plant tissue, and by saprophytic survival on these; (2) by saprophytic survival of the pathogen in dead infected tissues of a host plant, first invaded by the pathogen during its parasitic phase; (3) by dormant survival in the form of resting bodies (spores or sclerotia), which have been produced either in plant tissues colonized saprophytically, as under (1) above, or in tissues infected by the pathogen during its parasitic phase.

Because of the complexity of saprophytic competition for substrates amongst soil fungi, the circumstances under which some root-infecting fungi appear to have a phase of existence as competitive saprophytes in the soil are still incompletely known and imperfectly understood. In the earlier years of root-disease investigation, chance observations presented some valuable clues to a few gifted observers who were alert enough to be thinking about this problem already. The evidence for the conclusion that some tropical species of *Rosellinia* have a phase of active saprophytic colonization of lead litter on the forest or plantation floor; Botanist described the chain of evidence that permits us to conclude that *Rhizocrtonia solani* can exist in many cultivated soils as a free-living competitive saprophyte, for which it is well equipped by its ability to decompose cellulose quite vigorously. Since the beginning of the present century, mycologists from E.J. Butler (1907) onwards have been accustomed to investigate the occurrence of particular groups of fungi in the soil by burying therein 'baits' of particular types of dead plant or animal tissue so selected as to encourage competitive saprophytic colonization by the desired group of fungi, e.g. boiled hemp seed for *Pythium* species, human hair for keratin-decomposing fungi, shrimp or lobster shell for chitin-decomposing fungi, and so on. Sadasivan (1939) baited various soils at the Rothamsted Experimental Station with pieces of fresh, mature wheat straw, and in this way was able to confirm a conclusion from earlier circumstantial evidence that *Fusarium roseum* f. *cerealis* (syn. *F. culmorum*) is a regular primary fungal colonizer of this substrate when it is introduced into cultivated soils.

Isolation of fungi from such baits of dead plant or animal tissue buried in natural and otherwise untreated soil provides the most direct evidence obtainable that such fungi can colonize these particular substrates in soil as competitive saprophytes; for root-infecting fungi, it constitutes evidence for the possibility of an independent saprophytic phase in the soil. This realization provided the basis for a technique for comparing degrees of competitive saprophytic ability amongst root-infecting fungi that was devised and developed by several scientists at the Cambridge Botany School from 1950 onwards; it has been employed by workers elsewhere under the designation of the 'Cambridge Method'. Like many other experimental methods, this one purports to create a model of the natural situation or situations in the soil, and so invites questions as to its veracity. Two questions in particular need answering. First, do substrates of the type employed experimentally occur at all frequently in the soil. Secondly, when soil artificially inoculated with a particular fungus is employed, does the range of inoculum dosage correspond with the range of population level at which the fungus occurs naturally in cultivated soils? This second question arises from the early discovery that the inoculum potential of a fungus at the surface of a substrate partly determines the share of the substrate that it is able to colonize and exploit.

For these reasons, interpretation of experimental findings on saprophytic behaviour of root-infecting fungi has been hampered by doubts concerning the validity of the experimental models of a kind that has scarcely vexed investigation of other phases of the life-cycle. Thus the parasitic phase, saprophyticmycelial survival in dead, infected host tissues, and survival of dormant propagules all present problems that are much less complicated than that of saprophytic competition, so that no one has really needed to ask whether the experimental models are satisfactory replicas of the natural situation.

The opportunities for purely saprophytic colonization of the root system are probably very restricted by comparison with the volume of root tissue that is infected during the parasitic phase. Infection of living tissues continues into, or may even be initiated during, senescence whether natural or infection induced; it is finally halted only when tissue-resistance to invasion has declined to a level at which any obligate saprophyte can invade and occupy the remaining volume of root tissue. Living root tissues are reserved for the parasites, whereas dead tissues are open to invasion by all soil fungi; both in time and in space, therefore, opportunities for parasitic invasion by root-infecting fungi greatly exceed their opportunities for saprophytic colonization, and this difference is greatest for the most specialized parasites, because these are the ones least effectively restricted by host resistance during their parasitic phase.

So far as occupation of the root system is concerned, therefore, we can conclude that saprophytic colonization by root-infecting fungi is of minor importance by comparison with infection of living tissues and that for the specialized pathogens it is of negligible importance. But shoot systems of plants eventually reach the soil, and are indeed regularly incorporated therein in the normal course of agriculture; what opportunities does this give for saprophytic colonization by root-infecting fungi as an additional means of multiplication and survival? Hudson (1968) has recently produced a very timely and comprehensive review of investigations conducted, particularly during the last two decades, into colonization of plant shoot systems by air-borne fungi. From this it appears that much of the plant shoot system is colonized by fungi whilst it is still erect; specialized parasites invade the tissues whilst they are still in their prime, weak parasites follow during the phase of senescence, and even saprophytes can enter and colonize before the shoot system reaches the ground. This air-borne fungal colonization will begin earlier and extend further in a moist than in a dry climate, and its extent will increase with the time that the plant shoot system is allowed to remain standing. In a warm, dry climate or season, for instance, the straw of a wheat crop that is ploughed in before the first rains of autumn may consist largely of virgin, undecomposed plant tissue; in a moist climate or season and where ploughing has been delayed, the value of the straw as a substrate for soil fungi will have been correspondingly reduced by prior fungal colonization. It is important to keep in mind this limitation upon the availability of substrates for soil fungi, before we proceed to discuss what factors are likely to determine success in saprophytic competition. For any particular fungus, the share of a potential substrate that it can obtain will be determined: (1) directly by its *competitive saprophytic ability*, which is an intrinsic characteristic of the fungal species; (2) directly by its inoculum potential at the surface of the substrate; (3) inversely by the inoculum potential of its competitors.

## COMPETITIVE SAPROPHYTIC ABILITY

The definition of this term proposed in 1950 is 'the summation of physiological characteristics that make for success in competitive colonization of dead organic substrates'. A definition for a new concept has something in common with a specific binomial for a newly discovered organism; the less one knows about it, the more important it is to be able to identify it. This definition is chiefly revealing about the state of ignorance concerning competitive saprophytic ability at that time. Since then, research has clearly demonstrated that the possession of competitive saprophytic ability is not, as scientist had originally visualized it, the key to success in competitive colonization of substrates of all kinds; instead, they are beginning to realize that the kinds of competitive saprophytic ability are almost as diverse as the types of substrate open to saprophytic colonization. An obvious parallel is the diversity of athletic prowess that makes for human success in athletic competitions; a champion weight-lifter is unlikely to excell in short-distance sprinting and vice versa. A closer analogy for plant pathologists is to be found in the term 'pathogenicity', which is host-specific, and most of all so in specialized parasites. Just so, competitive saprophytic ability is substrate-specific. In the case of obligately saprophytic fungi, competitive saprophytic ability for the colonization of one or more types of substrate can be equated quite precisely with the general fitness of the fungal species for existence in one or more ecological niches, and the term thus appears to be redundant. But in the case of root-infecting parasites, for which the scientist proposed this term in the first place, it may be conceded a certain usefulness for the answer of certain questions such as the following. Can this fungus exist as a free-living saprophyte in the absence of host plants? If so, how well can it compete with obligate sapropytes? Is the possession of competitive saprophytic ability incompatible with specialization to a parasitic existence? The evidence to be presented later will show that we have made some progress towards an answer to these various questions.

To forestall possible confusion, it is now necessary to point out that the term 'substrate' is widely used in two senses, through the particular sense is usually evident from the context: (1) in a wide sense, to denote a corpus of material, usually of plant tissue, which a fungus has colonized and on which it is subsisting; (2) in a narrow sense, to indicate a particular constituent of plant tissue that is being decomposed by a particular fungus, e.g. a sugar, pectic substance, hemicellulose, cellulose or lignin. This latter sense is similar to that in which the term substrate for an enzyme is employed. When exploiting a corpus of plant tissue as a substrate (in the wide sense), a fungus may be utilizing several constituents of the tissue (i.e. substrates in the narrow sense) simultaneously; the biochemical versatility of the whole organism is wider than that of any one of its enzyme-sequences. 'Saprophytic sugar fungi' have been defined as those able to utillize only sugars and carbon constituents of plant tissues that are simpler than cellulose, i.e. they are incapable of degrading cellulose or lignin. Such a fungus may have a high degree of competitive saprophytic ability for colonization and exploitation of those carbon compounds that it can decompose but will clearly have none for cellulose or lignin, which it cannot degrade even in pure culture. Conversely, a celluiose-decomposing fungus with a high degree of competitive saprophytic ability for this substrate may succeed in securing only a small proportion of the sugars and simpler carbon compounds present in a corpus of fresh plant tissue buried in soil, because its growth rate is lower than that of the fast-growing saprophytic sugar fungi which are usually the pioneer colonizers; they get there first and may consume most of the simpler carbon compounds before slower growing cellulose-decomposers can establish themselves on the substrate.

It is certainly possible to compile a list, of all the physiological and biochemical attributes that could conceivably contribute to competitive saprophytic ability of one kind or another; few successful saprophytes will need, or are likely, to possess all these attributes in high degree, and the particular spectrum of attributes will be determined by the substrate. A short list of the more obvious attributes might be as follows:

(1) rapid germination of fungal propagules and speedy growth of young hyphae when stimulated by soluble nutrients diffusing from a substrate;

(2) appropriate enzyme equipment for decomposition of the more resistant carbon constituents of plant tissues, such as cellulose and lignin;

(3) excretion of fungistatic and bacteriostatic growth products, including antibiotics;

(4) tolerance of fungistatic substances produced by other soil micro-organisms.

The term 'antibiotic' is generally reserved for microbial growth-products active against other micro-organisms at a concentration of 10 ppm or less, and somewhat species-specific in their growth-stasis effects. The term 'fungistatic substance' is usually used with reference to a microbial growth-product that is active only at a significantly higher concentration that the minimum for antibiotic action, and it is non-specific in its effects; examples are carbon dioxide, the bicarbonate ion, and substances generating extremes of pH in the ambient solution.

We will briefly indicate the way in which these various attributes of a fungal species contributing to its competitive saprophytic ability may be distributed amongst various substrate-groups of soil fungi, in the form of a tabulation as shown:

| | *Speedy germination and growth* | *Enzymes for cellulose and/or lignin degradation* | *Production of antibiotics etc.* | *Tolerance of antibiotics etc.* |
|---|---|---|---|---|
| Primary saprophytic sugar fungi, as pioneer colonizers of fresh plant-tissue substrates | + | – | – | – |
| Secondary saprophytic sugar fungi, associated with cellulose and lignin decomposers | – | – | + | + |
| Root-surface and rhozisphere fungi | – | – | + | + |
| Cellulose and lignin decomposers | – | + | + | + |

Some of the evidence for the conclusions summarized in this tabulation will be presented later in this chapter, but a brief commentary will be helpful at this stage. The primary saprophytic sugar fungi are, by definition, pioneer colonizers, and they occupy their substrate ahead of competitors so that neither production of nor tolerance to antibiotic substances is likely to give them any significant competitive advantage. The other three substrate-groups live in an environment in which they are exposed to the full intensity of microbial competition, so we can expect that both production of and tolerance to antibiotic substances will confer an additional degree of competitive saprophytic ability.

Secondary saprophytic sugar fungi are designated as those that live as commensals with cellulose-and lignin-decomposing fungi, i.e. on the cellobiose and glucose released by hydrolysis of cellulose, and on soluble products of lignin breakdown. Arising from a particular interest in the utilization of cellulose during saprophytic survival of cereal foot-rot fungi in infected wheat straw scientist postulated the existence of this commensal group of secondary saprophytic sugar fungi. The occurrence of this type of commensalism has been most elegantly demonstrated by Tribe (1966). He was able to grow several species of *Pythium* (sugar fungi) on cellulose film in association with various species of cellulose-decomposing fungi; the amount of growth made by a species of *Pythium* in presence/absence of an associated cellulose decomposer was estimated quantitatively by oospore counts.

Commensalism of this type has also been demonstrated by Chang (1967) for another non-celluloytic fungus, *Humicola lanuginosa*, when grown in association with the strongly cellulolytic *Chaetomium thermophile* on filter-paper cellulose. When *C. thermophile* was grown by itself, soluble products of cellulolysis accumulated in the filter-paper pads, and the presence of reducing sugars was demonstrated; when grown in association with *H. lanuginosa*, weight of solutes was reduced, and only a trace of reducing sugars could be detected. This and other evidence on the ecological niche occupied by secondary saprophytic sugar fungi has been recently discussed by Hudson (1968).

## INOCULUM POTENTIAL

We have already defined and discussed inoculum potential as a determinant of infection. With a slight change in working, this definition can be made to apply equally well to competitive saprophytic colonization: *inoculum potential is defined as the energy of growth of a fungus available for colonization of a substrate at the surface of the substrate to be colonized*. It is perfectly true that the population level of a particular fungus in natural soil, or is inoculum dosage in an experiment, is one of the most important factors determining the share of a substrate that will be obtained by that fungal species in competition with others. But it is not the only factor, because the success of a population of fungal propagules in competitive saprophytic colonization will also be affected by other factors, both endogenous and exogenous. The growth vigour of the propagules will be determined partly by their age and nutrient status (endogenous factors) and partly by nutrients diffusing from the surface of the substrate and by environmental conditions (exogenous factors). This whole complex of factors affecting the energy of growth of a fungus at the surface of a substrate, and similarly that of its competitors, is summarized by the concept of inoculum potential, and so this concept is just as apt for research on saprophytic colonization as for research on the process of infection. In the broad sense of the term 'competition', fungi compete for a share of substrate not only with other fungi, but also with bacteria and actinomycetes, and at least with members of the soil micro-fauna amongst soil animals. Nevertheless, it is axiomatic in the science of ecology that the keenest competition occurs between organisms occupying the same or closely similar ecological niches; for this reason, no doubt, all the investigations to be discussed in this chapter have been restricted to competition between fungi, and therefore my discussion of microbial competition will be restricted to saprophytic competition between fungi.

Although we are using the concept of inoculum potential in order to elucidate the mechanism of saprophytic competition, we have to distinguish carefully between its role in this situation, and that as a determinant of root infection. In the process of infection, inoculum potential of the fungus is the force that opposes resistance of the host; only if the inoculum potential of the fungus is sufficient to overcome resistance of the host does a successful, progressive infection ensue. A substrate, on the other hand, is inert and offers no active resistance to colonization; the resistance to be overcome is constituted by the competition of other organisms. For this reason, the propagule-population (per unit area of host or substrate surface) component of inoculum potential appears to be relatively more important in competitive saprophytic colonization than it is in host infection. Perhaps we can make this more clear by an analogy with human behaviour in a situation all of us must have experienced. In some countries at all times, and in all at some times, there is disorderly behaviour amongst a crowd of intending passengers trying to get into a public-transport vehicle. The door is open, so the vehicle offers no 'resistance' to invasion; for any one intending passenger, the only resistance comes from the competition of others. But success does not necessarily go only to the strongest; a person who happens to be immediately opposite the open door when the vehicle comes to a halt may be able to board it virtually without opposition, provided always that the crowd accepts a certain minimum level of civilized behaviour. Thus an 'advantage of position' may more than compensate for physical weakness; just so, a high propagule-population of a particular fungal species at the surface of a substrate gives the species an advantage of position which can more than compensate for a low degree of competitive saprophytic ability. If thus favoured by a high population level, almost any fungus can obtain some share of a waiting substrate.

In conclusion, we can epitomize the mechanism of saprophytic competition between fungi in the following statement: *the share of a substrate obtained by any particular fungal species will be determined partly by its intrinsic competitive saprophytic ability and partly by the balance between its inoculum potential and that of competing species.* Environmental conditions in the soil, as we shall see, can strongly affect the outcome of saprophytic competition; they do this by altering the balance between inoculum potentials of the competing fungi. Saprophytic competition between fungi presents some of the most difficult problems in the whole field of root-disease investigation.

## THE CAMBRIDGE METHOD

The Cambridge method for investigating competitive saprophytic colonization of substrates by root-infecting fungi, has been developed in 1950 and employed by the following scientists F.C. Butler (1953), R.L. Lucas (1955), A.S. Rao (1959), R.C.F. Macer (1961) and R.L. Wastie (1961). Pure cultures of a fungus to be tested were grown upon a mixture of quartz sand + 3% maizemeal; when the culture was fully grown, it was easily miscible with soil in any desired proportion. In preparation for an experiment, the maizemeal-sand culture of the fungus was progressively diluted with an unsterilized soil of medium texture to give a dilution series; a series that we commonly employed contained the following percentages of pure-culture inoculum: 100 (pure culture alone), 98, 90, 50, 10, 2 and 0 (unsterilized soil alone). Several glass jars were filled with each of these mixtures; in each jar were then buried fifty standard *substrate units* of dead plant tissue. The substrate units were incubated in the inoculum-soil mixtures for about 4 weeks, so as to give the inoculant fungus ample time to colonize as many of them, in competition with other soil fungi, as it was capable of doing. At the end of this period, the number of substrate units colonized by the inoculant fungus in each of the inoculum-soil mixtures was determined.

For investigation by the Cambridge method, we selected four (subsequently five) cereal foot-rot fungi, because at that time most of my previous research experience had been with this group of pathogens. These fungi were *Fusarium roseum* f. *cerealis* (syn. *F. culmorum*), *Ophiobolus graminis, Cochliobolus sativus* (Ito & Kuribay.) Drechsl. ex Dastur (Stat. conid. = *Helminthosporium sativum* Pamm., King & Bakke), and *Curvularia ramosa* (Bain.) Boedijn. At that time, there was already good evidence for supposing that *F. roseum* was a vigorous competitive saprophyte and that *O. graminis* was a poor one; these two fungi thus provided reasonably reliable points of reference by which to judge the value of the method. The choice of this group of fungi had the further practical advantage that the host-plant tissue, which seemed the most appropriate to use in these tests, could be wheat straw, which was easily obtained and cut up into standard substrate units; short lengths (approx. 4 cm) were so cut from the long straw as to have a node at the lower end, and this served to hold the straw pieces together during subsequent handling. Before burial in the inoculum-soil mixtures, the straws were soaked in water or various nutrient solutions, and then autoclaved.

Various methods were tested for assessing the percentage of substrate units colonized by the inoculant fungus after the 4-week incubation period in the series of inoculum-soil mixtures. Washing, followed by surface-sterilization and planting out on nutrient agar, was found to be a reliable method only for fungi that have a high degree of competitive saprophytic ability for colonization of nutrient agar, like *f. roseum*; fungi with a low degree of competitive saprophytic ability for this substrate are suppressed by competitors. A perfectly reliable, though laborious, method was the wheat seedling test scientists had originally devised for determining longevity of saprophytic survival by *O. graminis* in similar pieces of infected or colonized wheat straw. A wheat seed is placed inside the lumen of each piece of wheat straw to be tested for presence of viable *O. graminis*; a number of such inseminated wheat straws are then planted together in a box of sand, which provides a rooting medium optimum for infection by *O. graminis*. Viable *O. graminis* is subsequently demonstrated by the occurrence of runner hyphae and lesions on the roots of the wheat seedlings after approx. 3 weeks; the method is suitable also for detection of other cereal root-rot fungi. A second reliable method was the 'sand plate' devised by F.C. Butler (1953) by which wheat straws were recovered from the inoculum-soil mixtures, washed and plated out on a layer of moist sand in Petri dishes; Macer (1961) subsequently found that surface-sterilization after washing the straw was not only unnecessary but was actually harmful in partially suppressing the inoculant fungus, and so was best omitted. The sand plate method is merely a variant of the 'moist chamber' method long employed by mycologists to encourage fungi to sporulate and thus reveal their presence within infected or colonized plant tissue. Sporulation on the surface of a colonized substrate within a moist chamber does not involve any appreciable degree of competition between fungi *already* established on the substrate, and so the method does not have the disadvantage of strong selectivity, as does plating out on nutrient agar. In is most satisfactory for fungi like *C. rativus* and *C. ramosa* that produce large, dark coloured spores that are easily visible under the stereoscopic microscope.

The results obtained by the Cambridge method for these four fungi can be illustrated by a selection from the data of F.C. Butler (1953). Butler tried all three methods described above for assessing percentage of straws colonized by his inoculant fungi. The data have been selected for each fungus from an experiment in which assessment of colonization was made by the wheat seedling test. The dilution end-point is likely to be lower for infection of wheat seedling roots than for sporulation in a moist chamber, so it is justifiable to compare only results obtained by the same technique, as I have done in Table 3.1.

**TABLE 3.1**

**Saprophytic Colonization of Wheat Straw by Four Cereal Foot-rot Fungi**

| | *Percentage straws colonized* | | | | | | |
|---|---|---|---|---|---|---|---|
| Percentage inoculum | 100 | 98 | 90 | 50 | 10 | 2 | 0 |
| *Fusarium roseum* | 90 | 87 | 80 | 67 | 65 | 55 | 22* |
| *Curvularia ramosa* | 100 | 96 | 96 | 96 | 85 | 88 | 0** |
| *Cochliobolus sativus* | 100 | 33 | 46 | 10 | 8 | 2 | 0 |
| *Ophiobolus graminis* | 98 | 46 | 8 | 10 | 2 | 0 | 0 |

* Colonized by *F. roseum* propagules occurring in the soil.
** *C. romosa* does not occur naturally in English soils.

As scientist had expected to find, the percentage of straws colonized by each inoculant fungus declined progressively as the inoculum was diluted with unsterilized soil; this not only diluted the inoculum but provided an increasing degree of competition from soil fungi and other micro-organisms with the increasing proportion of unsterilized soil. But as they had hoped to find, the *rate* of this decline with dilution of the inoculum provided an estimate of the degree of competitive saprophytic ability in the four inoculant fungi under comparison. Decline in percentage straws colonized by *F. roseum* and *C. ramosa*, respectively, was quite shallow down the dilution series; in contrast, decline by *O. graminis* and *C. sativus*, respectively, was comparatively steep. They thus felt justified in concluding that the first group, possessed a high degree of competitive saprophytic ability for colonization of wheat straw, whereas the second group possessed a much lower degree, at least under the conditions of these tests. These results for *F. roseum* and *O. graminis*, respectively, agreed with expectation from earlier evidence on their competitive saprophytic abilities.

Butler (1953) gave the results of various experiments designed to elucidate the difference in competitive saprophytica ability between *C. sativus* and *C. ramosa*. These two fungi had a growth rate over the surface of nutrient agar of a comparable order. Tests for antibiotic production by the stronger saprophyte, *C. romosa*, were negative, whereas both isolates of *C. sativus* produced antifungal culture filtrates; this difference was thus the reverse of that required to explain the difference in their respective competitive saprophytic abilities. But a possible explanation was furnished by the finding that *C. sativus* was much more sensitive to antifungal antibiotics, of which ten were tested, than was *C. ramosa; C. sativus* was also more sensitive to antagonism by some common soil bacteria in culture.

A fortunate discovery by Lucas (1955) led to an important new insight into the nature of saprophytic competition during the invasion of wheat straw tissue. Quite by chance, he discovered that, in straws from all the inoculum-soil mixtures he tested, *O. graminis* was more abundant in the outer than in the inner tissues of the straw. He interpreted this observation by supposing that the struggle for colonization of the straw between *O. graminis* and other soil micro-organisms in the soil component of the mixture was not settled once and for all in the outer layer of straw tissue; competition was postulated as continuing just as intensely whilst the organisms progressed inwards with the result that *O. graminis*, as a fungus of low competitive saprophytic ability, got left behind. If this supposition were correct, then one should predict that a vigorous competitive saprophyte like *F. roseum* should not get left behind in this way but should be as abundant within the inner as in the outer tissues. This point was verified by Macer (1961) and so these observations have confirmed the difference in saprophytic ability between these two fungi in another way. This finding led Macer to determine respective rates of straw tissue penetration in pure culture by these four fungi, together with a fifth, *Cercosporella herpotrichoides* Fron, which causes the eyespot lodging disease of cereals. The difference in straw penetration rates by *O. graminis* and *F. roseum*, respectively, turned out to be greater than that between their respective surface growth rates over nutrient agar.

Sufficient evidence has now accumulated for an attempt to interpret the marked difference in competitive saprophytic ability for straw tissue colonization between these two pairs of fungi: the vigorous saprophytes, *F. roseum* and *C. ramosa*, and the weak saprophytes, *O. graminis* and *C. sativus*. The data assembled for this purpose in Table 3.2 are: (a) the approximate percentage of unsterilized soil in the inoculum-soil mixture that permits 50% of the straws to be colonized by each inoculant fungus (Table 3.1); this gives a single-figure estimate (the ED50) for competitive saprophytic colonization; (b) superficial growth rates of the four fungi over nutrient agar, taken from Macer (1961); (c) figures for straw penetration rate in arbitrary units, taken from Macer (1961); (d) *a reduction factor* for each fungus, expressing the proportion by which growth is likely to be reduced by antagonistic effects from competing micro-organisms. These values of the reduction factor for each of the four fungi have been taken from results obtained with a cellophane-covered agar plate test devised by Wastie (1961).

It can be seen in Table 3.2 that there is no correlation between saprophytic success (a) and growth rate over nutrient agar (b). Straw penetration rate (c) is virtually the same for three of these fungi, but *O. graminis* is handicapped by a straw penetration rate only about 0.6 that of the others. But if values for straw penetration rate (c) are adjusted by multiplying by the reduction factor (d), to allow for the respective proportions by which growth rate through the straw tissue is likely to be reduced by fungistatic effects from other soil micro-organisms, then we get the product (cd) shown in the fifth column of Table 3.2. Mean values of cd for the two groups of vigorous and weak saprophytes, respectively, are correlated well enough with mean values for saprophytic performance. We can conclude that the poor performance of *C. sativus* under these experimental conditions can be ascribed chiefly to its low tolerance of fungistatic growth products produced by soil micro-organisms: this affords independent confirmation, from Wastie's values for the reduction factor, of Butler's (1953) original conclusion. The poorest saprophyte,

*O. graminis*, is handicapped both by low tolerance of fungistats and by a straw penetration rate lower than that of any of the other three fungi.

**TABLE 3.2**

**Distribution of Characteristics Possibily Determining Competitive Saprophytic Ability for Colonization of Wheat Straw Tissue Amongst Four Cereal Foot-rot Fungi**

| | *(a)* Percent. unster. soil permitting colonization of 50% straws by inoc. fungus at lab. temp. (16–22°C) | *(b)* Growth rate over agar at 22.5° C (mm/24 h) | *(c)* Straw penetration rate at 22.5°C (arbitrary units) | *(d)* Reduction factor due to competitors | *(cd)* |
|---|---|---|---|---|---|
| Fusarium roseum | approx. 98 | 10.8 | 10.0 | 0.87 | 8.7 |
| Curvularia ramosa | >98 | 4.8 | 10.4 | 0.51 | 5.3 |
| Cochliobolus sativus | approx. 10 | 3.8 | 10.3 | 0.16 | 1.6 |
| Ophiobolus graminis | approx. 2 | 8.0 | 6.4 | 0.24 | 1.5 |

These experiments at Cambridge were carried out under standardized conditions, so as to be comparable with one another, and the jars of inoculum-soil mixtures were incubated at laboratory temperature (range 16–22°C). But it has been pointed out that the share of substrate obtained by any particular fungal species would be determined by the balance between its inoculum potential and that of its competitors; we should further expect that this balance will be altered by a sufficient change in environmental conditions, which are one of the determinants of inoculum potential. This expectation has been confirmed by the work of Burgess & Griffin (1967) at the University of Sydney, employing the Cambridge method with two of our fungi, *Cochliobolus sativus* and *Fusarium roseum* f. cerealis, and two others, *Gibberella zeae* and *Cochliobolus spicifer* Nelson. Their experiments were designed to investigate the effect of temperature upon competitive saprophytic colonization by these four cereal foot-rot fungi; the temperatures they employed were 10, 20 and 30°C, together with two programmed régimes of fluctuating temperature, $10 \pm 5°C$ and $30 \pm 8°C$. All four fungi colonized the maximum percentage of wheat straws at the lowest temperature, 10°C. Their increased success at 10°C cannot be ascribed to a direct effect of temperature on their own inoculum potential, because for all of them the optimum for growth in pure culture was 25°C or above; instead, as Burgess & Griffin have suggested, we must conclude that a soil temperature of 10°C depressed the overall inoculum potential of competing saprophytes more than it depressed that of any of the four fungi under test. This situation thus closely parallels the effect of temperature upon pre-emergence killing of seedlings, elucidated by L.D. Leach (1947) although the optimum temperature for growth of the fungal pathogens is around 25°C, percentage pre-emergence killing in some host-pathogen combinations may be greatest at a much lower temperature, because the decrease in temperature retards velocity of seedling emergence more than it retards fungal activity.

In addition to reporting this important and hitherto undescribed effect of temperature upon competitive saprophytic colonization, Burgess & Griffin have also pointed out that at 20°C, approximating to the mean laboratory temperature at which Butler's (1953) experiments were incubated, they obtained higher percentages of saprophytic colonization by *Cochliobolus sativus* than Butler did. They suggest that this can be explained by supposing that the intensity of microbial competition was lower in their soil than in that employed by Butler. Burgess & Griffin (1968) reported natural colonization by *C. sativus* of wheat straw buried in two uninoculated soils and incubated at 10°C, although the spore population of *C. sativus* in both soils was too low to cause infection of wheat seedlings. They concluded that, at this temperature, *C. sativus* might be able to increase its population by saprophytic activity, and thereby attain an inoculum potential high enough for infection of wheat seedlings; their suggestion that wheat stubble should be ploughed into the soil in summer, whilst soil temperatures are still high enough to suppress saprophytic activity of *C. sativus*, seems a logical application of their findings to farm practice. A similar effect of temperature upon saprophytic activity of *Ophiobolus graminis* has been found by Gerlagh (1968), also employing the Cambridge method; percentage saprophytic colonization of wheat straws was found to increase down the temperature series 23, 17, and 10°C.

Whereas the results of Burgess & Griffin and of Gerlagh show that the original studies at Cambridge, made over a temperature range of 16–22°C, underestimated the competitive saprophytic activity of cereal foot-rot fungi under a lower temperature régime in the field, the work of Cook

& Bruehl (1968) in the pacific Northwest region of the USA suggests that another characteristic of the Cambridge method has probably overestimated it; these two opposite effects may to some extent cancel each other out. Selecting clean, bright unsterilized wheat straw for burial in containers of soil held at optimum moisture content and at 20°C, Cook & Bruehl found that 94–8% of straws were colonized by *Fusarium roseum* f. *cerealis* from propagules occurring naturally in the soil, which was not inoculated; determination of *F. roseum* was made by plating out straws on nutrient agar, a method quite appropriate for this particular fungus. Their results from this laboratory test thus repeated the same test in Cambridge. But when they made a survey of the fungal population in old straw pieces recovered 2–12 months after ploughing or cultivating into soil in the Pacific Northwest region of the USA, they found that the pattern of colonization of the straw pieces by *F. roseum* closely resembled the pattern of *infection* of the wheat plant by this fungus around harvest time, i.e., *F. roseum* was isolated most frequently from the foot or collar region of the straw and frequency of isolation declined with increasing height above soil level. Cook & Bruehl have therefore concluded that, in their wheat-growing area at least, multiplication of *F. roseum* in its parasitic phase much exceeds its increase in population by competitive saprophytic colonization of wheat straw; this conclusion accords with the more general view we expressed in the preamble to this chapter. These authors have discussed possible reasons for this apparent discrepancy between the results from the laboratory test and the field survey: (1) use of clean, bright straw for the laboratory test, whereas much of the straw ploughed into the soil after harvest may be 'weathered', i.e. precolonized by air-borne fungi; (2) planting of straws within a month of burial in the laboratory test but sometimes longer after ploughing in the field survey; thus Walker (1941) obtained a decreasing percentage isolation of *F. roseum* with time from first burial of straws in soil over the period 4–20 weeks tested; (3) employment of optimum soil moisture content in the laboratory containers, as compared with a drier surface soil after harvest in the field. In a survey of eighty randomly selected fields of wheat, Cook & Bruehl were unable to detect *F. roseum* by planting on a selective agar in fifty-eight fields, and in another thirteen fields the population was < 100 propagules/g soil; according to Cook (1968) these propagules are mainly chlamydospores. They argue that if *F. roseum* were a regular saprophytic colonizer of wheat straw in their area, then a higher and more consistent population level of this fungus in the soil would be found. From the survey made by Snyder & Nash (1968) of populations of *Fusarium* species in the soil of various fields at the Rothamsted Experimental Station, there is no doubt that frequent cropping with cereals does fovour the multiplication of *F. roseum* f. *cerealis*. Thus populations of $2-3 \times 10^3$ propagules/g soil were characteristic of Broadbalk Field cropped continuously with wheat since 1843, whereas from the soil of Barnfield, under root crops alone since 1876, *F. roseum* was rarely isolated. But what proportion of the population on Broadbalk has come from saprophytic colonization still remains a matter for conjecture.

Thus far we have endeavoured to interpret the available evidence on competitive saprophytic ability for colonization of a wheat straw substrate in terms of speed of propagule germination and fungal growth, production of fungistatic substances and tolerance of those produced by other soil micro-organisms. It is also possible that the rate at which a fungus can decompose cellulose and other more resistant components of wheat straw tissue may be a component of competitive saprophytic ability for this substrate, though this possibility has not been explored. Another possible component of competitive saprophytic ability has been suggested by Byther (1965) and Lindsey (1965) at the University of Colorado. Thus Byther found that *Fusarium roseum* was superior to *F. solani* in its utilization of inorganic nitrogen, and especially of nitrate nitrogen, in pure culture under sub-optimal conditions for growth, such as a low level of carbon nutrient, a high pH value of the growth medium and a low ambient temperature. Lindsey employed this information to interpret his data on saprophytic competition between *F. roseum* and *F. solani* for colonization of autoclaved soil and then for establishment in soil microbiological sampling tubes (Mueller & Durrell, 1957). *F. roseum* was the stronger saprophyte of the two in this experimental situation, but the degree of its dominance over *F. solani* was enhanced by shortage of available nitrogen, and reduced when nitrogen supply was not limiting growth. These results thus suggest that the speed and efficiency with which a fungus can absorb nutrients from a limited supply may be a factor contributing to its competitive saprophytic ability.

## COLONIZATION OF THE NUTRIENT AGAR PLATE

This subject has been of interest to two groups in particular amongst mycologists. First, mycologists studying the fungal flora of the soil have employed various types of nutrient agar as a substrate for the isolation of soil fungi. The techniques fall into two groups. In the first group, a soil suspension at a suitable dilution is incorporated with the melted agar to give a soil dilution plate, or finely sieved soil is mixed with melted agar in the Petri dish to give a 'soil plate'. In the second group, nutrient agar is exposed to colonization by soil fungi *in situ* in the soil, as in Chesters's (1940) soil immersion tube and its modifications, the screened immersion plate of Thornton (1952) and the soil microbiological sampling tube of Mueller & Durrell (1957). Secondly, all root-disease pathologists studying root-infecting fungi have frequent occasion to attempt isolation of their primary parasite, either for checking a disease diagnosis or for procuring a fresh isolate of the fungus for study. This isolation is often difficult of accomplishment without special precautions; specialized parasites more often than not have a low competitive saprophytic ability for colonization of the agar plate, and so they are suppressed by unspecialized parasites and saprophytes that have followed them into the root tissues. The special precautions that have to be taken for success include selection of recently invaded tissue and its surface-sterilization before plating out, both of which reduce the inoculum potential of competing fungi, and employment of a selective medium and/or temperature of incubation to favour the primary parasite over its fungal competitors. Competing bacteria are more easily suppressed, formerly by acidifying the agar and more recently by incorporation of antibacterial antibiotics with the plating medium.

It therefore occurred to scientists to modify the Cambridge method so as to make a more systematic study of competitive saprophytic colonization of the nutrient agar plate than had hitherto been attempted, and this was done in turn by Rao (1959) and Wastie (1961), employing a range of root-infecting fungi as subjects. In this agar-plate variant of the Cambridge method, the inoculum-soil mixtures were made up as before and sufficient aliquots of each to provide replicates were spread out as a layer 1.0–1.5 mm deep over the bottom of a Petri dish and then impregnated by and covered with water agar, first melted and then cooled down to approx. 40°C. After the agar had solidified, discs of 4 mm diam. were cut out with a cork borer and set out (four to a plate), on five replicate plates of modified Czapek-Dox + 0.05% yeast-extract agar. The number of pure colonies of the inoculant fungus arising from a total of twenty discs for each inoculum-soil mixture was then recorded. Partial colonies, i.e. a sector of the inoculant fungus in a composite colony with one or more other fungi, were recorded as fractions, the sum of which was then added to the number of pure colonies established by the inoculant fungus.

An improvement to the original Cambridge method in standardization of the pure-culture inoculum was introduced by Rao and employed by Wastie in all his final assessments of saprophytic performance on the agar plate by eleven of the root-infecting fungi that he tested. By the original method, pure cultures of all inoculant fungi on sand + 3% maizemeal were incubated at 25°C for about 4 weeks. We had realized that this standard period of about 4 weeks might not coincide with 'peak vigour' of all the fungi grown on the medium, and this period of incubation was fixed thus so as to be long enough for the slower-growing fungi. For all this eleven fungi, therefore, Wastie performed separate colonization experiments with pure cultures ranging in age (at 25°C) from 5 days by steps of 5 to at least 30 days. For his final results with each fungus, he then selected that age of the inoculum that had given maximum colonization of the agar plates by the inoculant fungus. In this way, it was possible to compare saprophytic performance by each fungus at its peak inoculum vigour, and so to eliminate one source of experimental error.

Wastie confined his experimental consideration of competitive saprophytic ability to the two characteristics of a fungus that then seemed most likely to determine its success on the agar plate: (1) growth rate in pure culture on the virgin nutrient agar; (2) tolerance of fungistatic microbial growth-products as expressed by the ratio 'growth rate on staled agar/growth rate on virgin agar'. For the purpose of determining these two values for each fungus, Wastie successfully developed an ancillary technique, the cellophane-agar-plate test. First of all, he inoculated an agar plate with soil at four equidistant loci, and then covered the whole plate with a thin sheet of wet cellophane. He then inoculated the surface of the cellophane, immediately above each site of soil inoculum, with an agar inoculum disc (4 mm diam.) of the fungus to be tested. In four series of replicate plates, the fungus was inoculated onto the cellophane at 0, 12, 24 and 48 h after the original inoculation of the plate with soil. By comparing rate of growth of the inoculant fungus over the cellophane on plates that had received no soil inoculum with those over plates receiving soil 0, 12, 24, and 48 h previously, Wastie was able to measure the depression in growth rate due to growth of soil fungi from soil inocula under the cellophane. This depression was ascribed to production of fungi-static growth-products by these fungi, accumulating in the agar and diffusing through the cellophane. In Wastie's Table 2, he has listed for fourteen root-infecting fungi the values of the ratio 'growth rate on 48 h staled agar/growth rate on virgin agar'. Under the name of Wastie's 'reduction factor', we have already made use of this ratio in Table 3.2 ; as

Wastie's values for all fourteen fungi are of general interest, we have reproduced them here in Table 3.3.

**TABLE 3.3**

**Tolerance of Fourteen root-infecting fungi towards fungistatic microbial growth-products, expressed by Wastie's 'reduction factor'**

| | | | |
|---|---|---|---|
| *Rhizoctonia solani* | 0.94 | *Macrophomina phaseoli* | 0.33 |
| *Fusarium roseum* f. *cerealis* | 0.87 | *Pythium mamillatum* | 0.25 |
| *F. oxysporum* f. *vasinfectum* | 0.80 | *Ophiobolus graminis* | 0.24 |
| *F. oxysporum* f. *cubense* | 0.66 | *Cochliobolus sativus* | 0.16 |
| *F. avenaceum* | 0.53 | *Verticillium dahliae* | 0.11 |
| *Curvularia ramosa* | 0.51 | *Helicobasidium purpureum* | 0.07 |
| *Fusarium caeruleum* | 0.35 | *Fomes annosus* | 0.06 |

Wastie obtained a complete set of the required data for eleven out of the fourteen fungi listed in Table 3.3. This comprised: (1) a colonization rating for each fungus at peak vigour of its inoculum; the rating was the mean percentage colonization for the whole range of inoculum-soil mixtures, which contained 98, 90, 75, 50 and 10% inoculum, respectively; (2) growth rate of each fungus in pure culture on cellophane overlying virgin agar, and growth rates across the cellophane when the underlying agar had been inoculated with soil 0, 12, 24, and 48 h previously. Wastie set out his final conclusions in the form of four correlation diagrams; we have given a summary of them in Table 3.4, more concisely and with a test of statistical significance, by means of correlation coefficients.

We can see in Table 3.4 that values of *r*, the correlation coefficient, fail to show any significant correlation between the colonization rating and either growth rate of the various fungi on 48 h staled agar or their tolerance of fungistatic growth-products. But the value of *r* for the correlation between saprophytic success and fungal growth rate on virgin agar is significant at the 2% level. This suggests that the outcome of the struggle between the inoculant fungus and other soil fungi in the inoculum-soil agar disc is decided before any fungistatic concentration of fungal growth-products has had time to accumulate in the agar; the issue may even be decided within the first 24 h after plating, because outwards hyphal growth from the inoculum-soil discs can usually be seen by the end of this period, at least under the microscope. If this assumption is correct, then Wastie would have obtained a still better correlation between saprophytic success and fungal growth rate if he had measured growth of his fungi in pure culture after the first 24 h. For nearly all the fungi that he tested, the first 24 h would have come within the lag phase of growth, i.e. the phase before the steady-state growth rate is attained. The lag phase is made up of the period before the fungus makes any microscopically visible growth together with the ensuing period over which growth rate is gradually increasing to the steady-state phase. For all his fungi except two very fast and one very slow grower, Wastie estimated growth rate by measuring colony diameters after 59 h. This value does not tell us enough about growth over the first 24 h, because the shape of the lag-phase part of the growth curve may vary widely between a number of fungi that have all attained the same colony diameter after 59 h.

TABLE 3.4

**Success of eleven root-infecting fungi in competitive saprophytic colonization of virgin nutrient agar from a series of inoculum-soil mixture**

| *Between colonization rating (arcsin-transformed) and the following:* | *Value of r* | *Value of p* | *Significance of r* |
|---|---|---|---|
| Tolerance of fungistatic growth-products (= reduction factor) | 0.2547 | > 0.1 | N.S. |
| Growth rate on 48 h staled agar | 0.5000 | 0.1 | N.S. |
| Growth rate on virgin agar | 0.6631 | 0.02 | At 2% level |

We can thus conclude from Wastie's data that for a primary saprophytic sugar fungus acting as a pioneer colonizer of a sugarbased nutrient agar, or of some kinds of dead but fresh plant tissue, the most important component of its competitive saprophytic ability consists in rapid germination and early growth of its propagules when activated by a nutrient stimulus from the substrate. The conclusion will not come as a surprise to any mycologist who has studied the soil fungus flora by one of the cultural isolation methods to which we have referred earlier in this section. In recent years, fungistatic substances like rose bengal have been added to nutrient agars employed for isolation of soil fungi, in order to slow down fast-growing fungi so that others get a better chance of developing colonies; this function of rose bengal is additional to its valuable bacteriostatic effect. Wastie's conclusion is also supported by the findings of Lindsey (1965), who inoculated pairs of fungi simultaneously into autoclaved soil, which was later assayed by the soil microbiological sampling tubes of Mueller & Durrell (1957); the faster-growing fungus of each pair was the one predominantly isolated by this procedure.

Fungal growth rate is also likely to be the paramount determinant in intraspecific fungal competition on the nutrient agar plate or slant tube. In a mass isolation of a fungal pathogen from infected tissue containing more than one wild-type strain of the fungus, the fastest-growing one is likely eventually to suppress the others in prolonged culture over a succession of transfers. This is well illustrated by Stover's (1950) study of the brown and gray wild types of *Thielaviopsis basicola*, causing black root-rot of tobacco. On potato-dextrose agar, the growth rate of the brown wild type was greater by approximately one-third than that of the gray, which tended to be dominated and eventually suppressed by the brown. In *Rhizoctonia solani*, virulence in unspecialized pathogens tends to be associated with a high mycelial growth rate; this association is found also in *T. basicola*, inasmuch as the faster-growing brown wild type is a more virulent pathogen than the gray. Prolonged culture on nutrient agar, with a succession of transfers to fresh media, does not usually result in loss of pathogenicity by stock cultures of unspecialized fungal pathogens; the procedure is selective for the fastest-growing strains in a mixture and therefore also for the most virulent ones. Unfortunately this mechanism operates in the reverse direction for the more highly specialized parasites, in which growth rate tends to be *inversely* correlated with the degree of parasitic specialization. Mycologists have found that stock cultures of *Ophiobolus graminis* are particularly liable to loss of pathogenicity in this way; no doubt many other mycologists have similarly lost both time and temper with other fungi. This difficulty in the maintenance of pathogenic wild-types has been much reduced by the modern practice of maintaining stock cultures in a dormant condition; agar cultures once grown can be covered with mineral oil or freeze-dried, or cultures can be grown on autoclaved soil that is then allowed to become air-dry.

### Competition on Staled Nutrient Agar, and Effects of Surface Sterilants

The conclusions concerning saprophytic colonization of virgin nutrient agar do not apply to colonization of the same agar when it has become staled by fungal growth, as scientist have demonstrated in a recent study at Cambridge. They found that the species-spectrum of fungi colonizing staled agar changed progressively with the time and degree of staling due to fungal growth; the most highly staled agar was colonized only by species of *Penicillium* and *Trichoderma*, which showed a high degree of tolerance to fungistatic growth-products of other fungi.

When plating out pieces of plant tissue for assay of their fungal colonizers, the type of surface-sterilizing agent employed and the duration of the treatment exert a selective effect on the species-spectrum of fungi forming colonies on the isolation medium. This was shown by Walker's (1941) study of fungi colonizing wheat straw baits buried in various soils collected from fields at the Rothamsted Experimental Station. *Fusarium roseum* f. *cerealis* was the dominant fungus on the isolation plates after treatment of the straws with calcium hypochlorite, a mild sterilant, or after prolonged washing with sterile water. But after treatment of the straws with mercuric chloride or silver nitrate, the place of *F. roseum* was taken by species of *Penicillium* and *Trichoderma* which, after the longer periods of surface sterilization, were often the only fungi developing on the isolation plates.

This effect has been put to good use in the search for surface sterilants that are selective for particular fungal pathogens. In an early application of this principle, Davies (1935) found that a higher percentage isolation of *Ophiobolus graminis* from infected wheat tissue could be obtained by employing silver nitrate instead of the more usual mercuric chloride, and we have found that this method gives better results than any other. Since then, this principle has been widely employed in the search for selective surface sterilants, and for selective fungistatic agents to be incorporated in the isolation media designed for picking up particular pathogens.

## COLONIZATION BY *RHIZOCTONIA SOLANI*

We must first mention, some most elegant work by Bateman (1964) on the decomposition of cellulose by *R. solani* in pure culture. He demonstrated that *R. solani* could decompose cotton fibres, which are generally agreed to be the form of native cellulose most resistant of any to decomposition, and so this fungus must be regarded as a highly efficient cellulose-decomposer. Bateman's paper is accompanied by some excellent photomicrographs showing enzymic erosion of cotton and filter-paper fibres taken from his pure cultures of *R. solani* on these two substrates. He identified areas of cell-wall hydrolysis by their loss of birefringence under polarized light, thus recalling the earlier studies by

F. Baker (1939) of cell-wall degradation in comminuted fragments of plant tissue by the rumen microfloras of various herbivores.

Much attention has been devoted to the saprophytic life of *R. solani* by G.C. Papavizas & C.B. Davey of the U.S.D.A. at Beltsville, Maryland. In their first papers on this subject (1961), they reported obtaining high percentages of *R. solani* from mature, dry stem segments of buckwheat (*Fagopyrum esculentum*) when buried in various soils for 4 days. Colonization of the baits by *R. solani* was determined by plating them out on water agar fortified with aureomycin, neomycin and streptomycin. If the baits were left in the soil for longer than 4 days, then percentage isolation of *R. solani* declined, presumably because of increasing competition for colonization of the agar from later fungal invaders of the bait tissues. Water agar is often employed instead of nutrient agar for attempted isolation of particular fungi from infected or colonized plant tissue, for the reason that both density of mycelial growth and production of fungistatic growth-products are restricted by the absence of nutrients, and particularly of carbon nutrients. But in view of Wastie's (1961) conclusion that success in fungal colonization of the virgin nutrient agar plate is determined chiefly by fungal growth rate, it is doubtful whether the precaution of using water agar in place of nutrient agar was of much help to Papavizas & Davey in isolating *R. solani* from their bait segments. For the particular case of *R. solani*, Wastie's data strongly suggest that its lack of success in colonization of the nutrient agar plate is due to the fact that it is not a fast grower, and even more to the fact that it is probably a late starter in the race. Thus out of eleven fungi for which full data were available, *R. solani* was ninth in percentage of colonies established on the plates, eighth in its growth rate on virgin agar, and first in its degree of tolerance to fungistatic microbial growth-products. This could be further confirmed from scientists experience in attempting to isolate *R. solani* from roots of young cereal plants suffering from the 'no growth' disease in South Australia; it was rarely possible to isolate the pathogen from diseased roots unless they were in the early stage of infection.

Papavizas & Davey obtained higher percentages of isolation of *R. solani* when their baits were buried at soil moisture contents below rather than above 50% saturation. Rather surprisingly, the dead buckwheat stem segments gave higher percentages of *R. solani* on plating out than did infected hypocotyls of snap bean (*Phaseolus vulgaris*). In a subsequent paper, Papavizas & Davey (1962) described some modifications in their plating out procedure for isolation of *R. solani* from mature stem segments (5–8 mm long) of twelve plant species tested as baits; the best results were given by buckwheat, cotton and lima bean and the poorest by maize, oats and soybean. Twenty clones of *R. solani* obtained from ten different soils were tested for pathogenicity against seedlings of snap bean, radish, soybean, sugar beet and wheat, and these saprophytically isolated clones were found to range from non-pathogenic to highly pathogenic; three clones were highly pathogenic to all five seedlings species. Percentage isolation of *R. solani* from these baits ranged from 1% from a soil under grass for 3 years up to 97% from a soil cropped intensively with susceptible species for at least 6 years. Following upon a demonstration by Papavizas & Davey (1962) that percentage saprophytic colonization of buckwheat baits by *R. solani* was progressively inhibited in four different soils by concentrations of carbon dioxide ranging from 10 through 20 to 30%, Papavizas (1964) studied carbon dioxide tolerance of twenty isolates each of *R. solani* and *Rhizoctonia praticola* Kotila, respectively. With a few exceptions, a high degree of competitive saprophytic ability amongst these forty isolates was found to be correlated with good tolerance of the higher concentrations of carbon dioxide. The results of Papavizas & Davey with stem segments as baits for *R. solani* have been confirmed by Zarka (1963), employing stem segments of *Corchorus olitorius* as substrate units in conjunction with the full Cambridge method; in his tests, the saprophytic performance of *R. solani* was as good as that we obtained with *Fusarium roseum* and *Curvularia ramosa* in colonizing wheat straw at Cambridge. Nevertheless, it is unsafe to generalize about the relative competitive saprophytic abilities of different fungal species from a comparison between single isolates of each; this has been demonstrated by the variation in saprophytic performance amongst the forty isolates of *R. solani* and *R. praticola* in the experiments by Papavizas (1964) just quoted, and has been further emphasized for *R. solani* by Baker, Flentje, Olsen Stretton (1967).

The findings of Papavizas & Davey in the U.S.A. have been very thoroughly confirmed by Sneh, Katan, Henis & Wahl (1966) in Israel, in a paper that provides some comparisons of particular relevance to this discussion. Sneh et. al. employed the method of Papavizas & Davey, and their isolation medium for plating the baits was water agar + 250 ppm chloramphenicol. As baits they employed stem segments taken from plants of bean (*Phaseolus vulgaris*), cotton, buckwheat, oats, wheat and barley; the bean and cotton plants were grown to an age of 6 weeks, whereas stems of the other four species of crop plant were described as 'mature'. No one acquainted with these six plant species would doubt that the greatest contrast in condition of the plant tissues would have been provided by the living green tissues of the 6-weeks-old bean and cotton plants, on the one hand, and the mature, dry and nearly dead tissues of wheat and barley on the other. We should therefore expect that the green tissues of bean and cotton would possess at least some residual resistance to invasion, thus favouring parasitic fungi like *R. solani* over obligate saprophytes, whereas the nearly dead mature tissues of wheat and barley would have lost any significant resistance to invasion by obligate saprophytes. Sneh et. al. have obligingly furnished us with a means of testing this assumption, by killing half their bait segments in the autoclave before burial in the soil. A relevant selection from their data is given in Table 3.5.

TABLE 3.5

**Percentages of untreated and autoclaved stem segments giving colonies of Rhizoctonia species on water-chloramphenicol agar after burial in a naturally infested soil for the periods shown**

| | | *No. days burial in soil* | | | | | |
|---|---|---|---|---|---|---|---|
| | | *2* | *3* | *4* | *7* | *14* | *21* |
| Bean, | untreated | 27 | 57 | 75 | 42 | 45 | 35 |
| | autoclaved | 46 | 57 | 56 | 27 | 18 | 25 |
| Cotton, | untreated | 24 | 64 | 70 | 66 | 45 | 45 |
| | autoclaved | 70 | 60 | 61 | 41 | 36 | 29 |
| Barley, | untreated | 51 | 47 | 38 | 7 | 0 | 0 |
| | autoclaved | 51 | 51 | 37 | 2 | 0 | 0 |
| Wheat, | untreated | 41 | 49 | 30 | 7 | 2 | 0 |
| | autoclaved | 50 | 51 | 30 | 6 | 0 | 0 |

We can see in Table 3.5 that percentage isolation of *Rhizoctonia* species (*sic*) from the bait segments has reached a higher maximum, and has continued at a higher level, in the untreated *living* stem segments of bean and cotton than in those *killed* by autoclaving before burial in the soil. This is exactly the result we should have expected; the living segments retain enough tissue-resistance to delay colonization by competing obligate saprophytes but not sufficient to delay infection by the parasitic *Rhizoctonia* species. Killing the stem segments by autoclaving has destroyed the advantage enjoyed by *Rhizoctonia* species of being able to invade the plant tissues in advance of obligate saprophytes. When we look at the corresponding figures for stem segments of wheat and barley, we see that killing the tissues by autoclaving has made no significant difference to percentage recovery of *Rhizoctonia* species on the isolation plates; it is clear that these mature tissues must have been virtually dead before autoclaving, so far as any resistance to colonization by obligate saprophytes was concerned.

This experiment by Sneh et. al. has been most admirably designed for a demonstration of this important distinction between infection of still living tissues by parasites and saprophytic colonization of dead tissues by obligate saprophytes. In any situation of disease risk, it is safer to employ dry, dead crop residues as an organic manure rather than the still living green tissues of a cover crop.

## COLONIZATION OF WOODY SUBSTRATES

The problems we have just been discussing have not been easy ones to investigate; nevertheless, if we now pass on to consider competitive saprophytic colonization of woody tissues, as provided by the old roots, branches and trunks of bushes and trees, we encounter situations of still greater difficulty. We have just been considering the importance of distinguishing between infection of still living tissues by parasites, and colonization of dead tissues by obligate saprophytes; this can be a difficult distinction to make in practice for the tissues of herbaceous plants, but with woody perennials the difficulty is formidable indeed. Yet some way o^f making this distinction has to be found, because reliable root-disease control in horticultural and sylvicultural practice can be securely based only on a thorough understanding of the fundamental issues involved.

Introduced by R. Leach (1939) in East Africa against *Armillaria mellea*. By the method, of "Ring-barking" old trees that have to be removed in preparation for establishment of a young plantation crop on the site are ring-barked a year or more before felling; the relatively early death of the root system that follows ring-barking permits earlier invasion of the tissues by saprophytic wood-decomposing fungi, and their occupancy brings to a halt further invasion by *A. mellea*. Prior saprophytic colonization thus provides a very useful control measure against the further spread of infection by specialized parasites like *A. mellea*, and so we need more information about this subject.

In an attempt to provide some information on this general question, Mycologists performed some laboratory experiments with *A. mellea* of a similar type to that by Sneh et. al. (1966) just described. As a substrate for

*A. mellea*, they employed segments of willow (*Salix alba* var. *coerulea*) shoots (approx. 1.75 cm diam.), which were buried in two types of soil in laboratory containers. Before burial, half of the living shoot segments were killed by autoclaving. Segments were buried in soil for various periods up to an effective maximum of 49 days before inoculation with *A. mellea* rhizomorphs growing from woody inocula. The relative volume of each substrate segment colonized by the fungus after an incubation period of 4 weeks was then assessed by measuring weekly growth increments of rhizomorphs put out from the segment in soil growth tubes over an assessment period of 5 weeks, employing a technique devised in an earlier study. The substrate value of the living segments for *A. mellea* was maintained for a period of 7 weeks previous burial in soil, which was the maximum tested. But the substrate value of dead segments declined with increasing length of burial in soil from 3 weeks onwards; this they attributed to partial occupation of the segments by saprophytic fungi before they were inoculated with *A. mellea*. Their results thus agreed with prediction from Leach's original hypothesis, though they had reduced the time-scale as well as the space dimensions of the experiment from those of the plantation crop to those of the laboratory.

The results of this experiment thus suggest that *A. mellea* can colonize dead woody tissues in the soil, or above it, if favourably placed to do so. Findlay (1951) has reported extensive damage by *A. mellea* to pit props in very damp coal mines. Other fungi infecting tree roots may similarly have opportunities for purely saprophytic invasion of woody tissues. Wallis & Reynolds (1965) have demonstrated that this is at least a possibility for *Poria weirii* Murr., causing a root disease of Douglas fir. Stakes of Douglas fir heartwood, air-dried for 6 months and thus 'presumed dead', were buried beneath a 50-year- old stand of Douglas fir in British columbia for periods of 0, 3, 6 and 12 months, respectively. Ten replicate stakes at each sampling period were removed from the soil, inoculated with wood-block inocula of *P. weirii* and then reburied. Saprophytic colonization of these stakes by *P. weirii* from the inocula did not decline with increasing periods of burial in the soil up to 6 months; after 12 months, however, there was a significant reduction in the extent of saprophytic colonization by *P. weirii*.

# 4

# ROOT-INFECTING FUNGI IN INFECTED OR COLONIZED HOST TISSUES

Saprophytic survival is defined as mycelial survival of a root-infecting fungus in infected host tissues invaded during its parasitic phase. We can also include in this term the mycelial survival of a parasite in plant tissues of any kinds that it has been able to colonize competitively in its saprophytic phase, if it has one. In respect of saprophytic survival, no valid distinction can be made between the mechanism of mycelial survival in infected and saprophytically colonized tissues, respectively; in fact, most of the investigations into a saprophytic survival in infected tissues have actually been carried out with colonized tissues. For such experiments, large numbers of infected or colonized tissue-units are required; infected material of uniform quality is difficult to collect from diseased crops in the field in sufficient quantity for a large experiment, and laborious and time-consuming to produce from inoculated plants in the glasshouse or in small field plots. But by inoculating prepared tissue-units after autoclaving in culture flasks, large quantities of material uniformly colonized by the desired pathogen can be prepared with the minimum of labour. Confirmatory tests have shown that such colonized tissue-units satisfactorily reproduce the behaviour of naturally infected tissue-units so far as saprophytic survival of the pathogen is concerned.

During infection of living tissues or saprophytic colonization of dead tissues, the fungus establishes a mycelial network throughout the tissues. Thereafter, whether the dead tissues remain in soil or are cultivated into it, the mycelial network of the pathogen survives saprophytically by a slow but continuous decomposition and exploitation of the tissues; sugars and the more easily assimilated compounds are quickly exhausted but the cellular framework of hemicellulose, cellulose and lignin is more resistant to decomposition, and provides for a considerable longevity of the pathogen; this longevity extends up to a year and sometimes more for pathogens of annual crops and for periods of years for pathogens of woody perennials. As we have remarked above, survival of a pathogen in inoculated and colonized tissues-units seems to reproduce satisfactorily, for experimental purposes, its survival in similar units of naturally infected tissue. In either case, the pathogen establishes its mycelial network in the plant tissue well in advance of other invaders; during infection, the pathogen invades living plant tissue, and host resistance keeps out secondary fungi, at least until the mycelium of the pathogen has completely occupied most, if not all, of the host tissue. Inoculation of sterilized tissue-units with the pathogen reproduces, reasonably well according to all available evidence, infection of living tissues for the production of survival-units. But when an unspecialized parasite with sufficient competitive saprophytic ability colonizes competitive dead crop residues that have been cultivated into the soil, it may not achieve so complete an occupation of the plant tissues as if it had infected them during its parasite phase; such incomplete occupation of the plant tissues is likely to be reflected in a decreased longevity of saprophytic survival.

Saprophytic survival of the mycelial network in plant tissues needs to be sharply distinguished from *dormant survival* of a pathogen, within the tissues or set free in the soil, in the form of resting spores or sclerotia. Such a distinction is easy to make in theory, but special precautions may have to be devised in experimental practice to separate these two types of survival, when they occur concurrently during the survival of a single fungus in plant tissues.

## SURVIVAL IN TISSUES OF HERBACEOUS HOSTS

Most of the investigations so far conducted into saprophytic survival have been made with the five cereal foot-rot fungi viz. *Ophiobolus graminis, Fusarium roseum* f. *cerealis* (syn. *F. culmorum*), *Cochliobolus sativus* (stat. conid. = *Helminthosporium sativum*), *Curvularia ramosa*, and *Cercosporella herpotrichoides*. The specialized ectotrophic parasite, *O. graminis*, infects first the root system, and thence the stem (tiller) bases of its cereal hosts; the root-rot thus develops into a foot-rot. There is less certainty about the usual sequence of infection in the foot-rots caused by *F. roseum, C. sativus* and *C. ramosa* but it seems probable that stem-base tissues are often invaded by these less specialized pathogens direct from the surrounding soil, and not necessarily via infection of the roots. The specialized pathogen *C. herpotrichoides*, causing the eyespot lodging disease of cereals, is not a root-infecting fungus at all. Primary infection of autumn-sown cereal seedlings occurs during the winter and early spring from rain-splashed spores; these spores are abundantly produced, when night temperatures are around 0° C, on infected stubble from a previous diseased cereal crop that has been incompletely buried during ploughing (Glynne, 1965). Although *C. herpotrichoides* does not infect the roots of its cereal hosts, the fungus is a soil-borne pathogen inasmuch as it survives saprophytically in infected cereal stubble and so its survival has been studied by the same techniques as have been employed for the other four fungi listed above. To these five fungi has recently been added a sixth; this is *Cephalosporium gramineum* Nisikado & Ikata (perfect state = *Hymenula cerealis* Ellis & Everh.), which causes a vascular wilt, known as stripe disease, of winter wheat in Washington State, U.S.A.

In an early studies of the saprophytic survival of *Ophiobolus graminis*, scientists employed a standardized survival-unit of wheat straw tissue colonized in pure culture by the fungus, though they took the precaution of confirming the more important conclusions resulting from such experiments by comparable tests with pieces of naturally infected wheat straw collected from diseased crops around harvest time. These survival-units were cut from lengths of mature wheat straw to give pieces about 4 cm long, each with a node at the lower end, so that the internodal part above it consisted of the culm surrounded by the subtending leaf sheath; inclusion of a node with each straw piece was essential to secure coherence during decomposition in the soil for prolonged periods, and during subsequent manipulation in the test for survival of viable *O. graminis*. After the straws had been colonized in pure culture by *O. graminis*, during incubation for about one month at 25° C, they were buried in lots of fifty in approx. 200 ml soil in glass tumblers or jars. At intervals, a sample of not less than 100 straws in two soil containers was taken for the survival tests; a wheat seed was planted, germ end downwards, within the lumen of each straw, and fifty straws thus inseminated were planted in a seedling box filled with sand, which is a rooting medium optimum for infection and spread by *O. graminis*. This wheat seedling test is the only reliable one for *O. graminis*, which is a poor competitor on the nutrient agar plate; the microconidia of this fungus are unreliable for diagnosis, and production of perithecia in a moist chamber cannot be induced either quickly or with the complete assurance required for a survival test. This difficulty does not arise with *Cochliobolus satius* or with *Curvularia ramosa*, which produce their large and distinctive, septate, dark-coloured conidia as soon as straws are incubated in a moist chamber of the sand-plate type devised by Butler (1953) and employed by him in subsequent studies of saprophytic survival. When straws have been incubated on a sand plate for about 1 week at 25° C, sporulating conidiophores on the straw surfaces can be recorded during rapid scanning under the stereoscopic microscope; if a detailed record is required, the complete length of the straw can be divided up into successive fields of view, each of 1 mm diam., by employing a mechanical stage. Survival of *Fusarium roseum* can be reliably assessed by planting out surface-sterilized straws on nutrient agar, because *F. roseum* has a high competitive saprophytic ability for this substrate. As for *Ophiobolus graminis*, survival of *F. roseum*, *C. sativus* and *C. ramosa* can be assessed by wheat seedling test; the test is sufficiently reliable for experiments in which straws have been completely colonized by the original inoculant fungus, and there is no substantial probability that the test wheat seedlings will be infected by other soil-borne fungal pathogens. This proviso has to be made, because roots infected by all three pathogens show a generalized discoloration due to cortical necrosis, and symptoms of root infection by one pathogen cannot be distinguished from those produced by the other two. This difficulty of recognition does not apply to roots infected by *Ophiobolus graminis*, which show two distinctive features: (1) an intense necrosis and discoloration of the vascular cylinder; (2) external growth of the characteristic, dark-coloured runner hyphae, which are easily visible under the stereoscopic microscope. The wheat seedling test for survival of *F. roseum* can be replaced either by plating out surface-sterilized straws on nutrient agar, as noted above, or by a modification of Butler's sand-plate test introduced by Macer (1961). The hyaline macroconidia of *F. roseum* are neither large nor distinctive enough to be identified with certainty under the stereoscopic microscope but they can be scraped off the straw surface into a drop of water, which is then searched under a sufficiently high power of the compound microscope. Macer employed this modified sand-plate method also for assessing survival in straw of *Cercosporella herpotrichoides*; to induce sporulation by this fungus, sand plates have to be subjected to a daily alternation of temperature with night temperature going down to near 0° C, as was first demonstrated by Glynne (1953). For *C. herpotrichoides*, Macer also successfully employed a wheat seedling test; nodes were cut off from straws after recovery from the soil, and the internodal segments were slipped over the coleoptiles of just emerged wheat seedlings as a collar. The inoculum segments were then covered with a layer of moist sand to ensure infection, the occurrence of which was demonstrated after some 8 weeks by appearance of the characteristic eyespot lesions.

During the first part of investigation into the saprophytic survival of *O. graminis*, it became possible to correlate the rapidity of disappearance of the fungus from the straws in a general way with the maintenance of soil conditions optimum for microbial activity; thus loss of viability was most rapid under conditions of medium to high temperature, suitable moisture content, and good soil aeration. Viability of *O. graminis* was preserved, on the other hand, at a low temperature (2–3° C) and in air-dry soil, and was actually maintained for longer in a water-logged soil than in one held at a moisture content of 50% saturation. The general interpretation placed upon these results is that under conditions conducive to maximum microbiological activity, the available food reserves of the straw substrate are most quickly exhausted by *O. graminis* itself and by associated micro-organisms that have followed it into the tissue. Conditions restricting microbiological activity, such as low temperature, dryness and poor aeration, are thus likely to promote maximum longevity of the fungus in infected host tissues, providing that the effect of such conditions is not so extreme as to be actually lethal. Thus *O. graminis* is likely to suffer less loss of viability during the low temperatures of winter, or during the extreme drought of such a summer as occurs in South Australia, than during periods of the year in which soil temperature and moisture content are more favourable for microbiological activity.

This conclusion has been fully confirmed by the work of Fellows (1941), using a technique approximating more closely to actual survival of *O. graminis* in the field. Fellows studied the rate of disappearance of *O. graminis* in naturally infested soil taken straight from the field and incubated under the eight possible combinations of high and low temperature, loose and firm packing, and moist and air-dry conditions. The low-temperature samples were stored in a deep cave and the high-temperature ones were kept in the glasshouse. Looseness was maintained by occasional stirring, and a firm packing achieved by thorough tamping of the soil whilst moist. The moist soils were maintained at approximately 60% of saturation, whereas the others were allowed to air-dry. Four experiments were made, and periods of incubation varied from 230 to

777 days; at the end of the incubation period, all soil samples were brought to favourable tilth and moisture content, and a test crop of wheat grown in them to maturity, in the glasshouse. The degree of infection on the plants at harvest was expressed by means of a 'severity rating'. In almost every comparison provided by this series of extensive factorial experiments, the fungus disappeared more quickly in a warm soil than in a cold one, and more quickly in a loose soil than in a compact soil. The influence of moisture was less straighforward; under cool conditions survival appeared to be rather better in a moist than in a dry soil, whereas under warm conditions the reverse was the case. In all four series of experiments, without exception, disappearance of the fungus was most rapid in the warm, moist, loose soil, i.e. under just those conditions of high temperature, adequate moisture content and good aeration postulated above as most conducive to rapid loss of viability.

**Effect of nitrogen on saprophytic survival**

The C/N ratio of mature wheat straw varies around a value of 80; mature stem tissues of other annual crop plants similarly have a fairly high C/N ratio. It was first realized many years ago, as recounted by Waksman (1932), that such mature plant tissues when composted were decomposed much more quickly if nitrogen, in some readily available form, was added to the compost; similarly, tissues of this type disappeared more rapidly in nitrogen-rich than in nitrogen-poor soils. The mechanism of this effect depends upon the fact that microbial degradation of pectins, hemicelluloses, celluloses and lignins of the plant cell wall is brought about by the action of enzymes. These enzymes are both bound to the microbial cell wall and set free in soil moisture films; the action of bound enzymes is necessarily very local and that of the free enzymes is limited by the number of enzyme molecules produced and by the effective radius of their diffusion. The largest constituent of mature plant tissues is cellulose and it is this, together with the superimposed lignin, that survives for longest in the decomposing tissue. From an active length of fungal hypha lying alongside a plane of plant cell wall, cellulase enzymes diffuse outwards and the products of cellulose hydrolysis, i.e. cellobiose and eventually glucose, diffuse inwards so that a proportion is absorbed by the fungal hypha. So the radius of cellulase action around the enzyme-producing fungal hypha must be limited both by the intensity and duration of its production, and also by any factors limiting its outwards diffusion. For these various reasons, no individual fungal hypha can produce more than a limited zone of enzymic erosion around itself; when this zone is exhausted, then this particular fungal hypha will die from carbon starvation.

Zones of enzymic erosion around fungal hyphae and bacterial cells have been demonstrated under the microscope by the use of differential staining methods and also by loss of birefringence of degraded cellulose when viewed by plane-polarized light. Both methods were used by F. Baker (1939), who has figured very clearly these zones of erosion around bacterial cells lying alongside cell walls in fragments of plant tissue taken from the alimentary canal of various herbivorous animals. Bateman (1964) has similarly figured them on cotton and filter-paper fibres taken from his pure cultures of *Rhizoctonia solani* on these two types of native cellulose. Enzymic erosion is also demonstrated particularly well by the so-called 'boreholes' around hyphae of wood-destroying fungi; Bailey & Vestal (1937) have used the geometrical configuration of these boreholes as an aid to interpreting the fine structure of cellulose fibrils. Proctor (1941) has provided microscopical evidence to show that penetration of the thickened secondary walls of plant xylem elements by fungal hyphae is through enzymic action and that mechanical pressure plays only a subsidiary role at most; he has suggested that the term 'borehole' is thus a misnomer. The observed fact that these boreholes do not usually attain a diameter exceeding three times that of the fungal hypha that has made them suggests that the functional life of any particular length of fungal hypha in cellulolysis is of limited duration. In decomposing cellulosic tissues of herbaceous plants, which are of a construction much looser than that of secondary wood in woody perennials, there must be a less effective barrier to outwards diffusion of both cellulase and the sugars derived from its action. The sugars will be quickly absorbed by associated micro-organisms living as commensals of the cellulose-decomposer in the plant tissue. We can therefore conclude that production of cellulase by a fungal hypha is likely to have become limited by the operation of natural selection to an optimum or economic quantity; we can certainly visualize the possibility that production of cellulase in excess of this optimum quantity would be uneconomic, because further increases in production would yield a progressively diminishing return of sugar by inwards diffusion towards the cellulase-producing hypha in the presence of sugar-absorbing commensal organisms. A way in which this limitation of cellulase-production may be brought about is suggested by the fact that the cellulases are *adaptive* (as distinct from *constitutive*) enzymes and so contact with cellulose is required for their induction; hydrolysis of the immediately surrounding cellulose will remove a fungal hypha from contact with its cellulose substrate, and so induction of cellulase will cease.

Nevertheless the mycelium of which our particular length of hypha under consideration forms a part can survive granted one essential condition; this is that a continued slow growth and extension of the mycelium within the substrate tissue is not limited by lack of nitrogen or of any other mineral nutrient. So we can now understand that, because both the working life and the survival of individual hyphae are limited by the radius of enzyme action, it must follow that a continue extension of the mycelium by branching and growth within the substrate tissue is an essential condition for continued cellulolysis, and also for survival of the mycelium effecting that cellulolysis. Rate of decomposition of plant tissue is thus limited by the supply of mineral nutrients essential for the synthesis of fungal and other microbial protoplasm. Nitrogen is the mineral element required in largest amount, followed by phosphate and potash. Decomposition-rate of mature plant tissue in compost or in nitrogen-poor soil is thus limited by the supply of available nitrogen; phosphate may sometimes limit decomposition-rate in compost, but much more rarely in the soil.

An understanding of the mechanism of microbial decomposition of plant tissue, as outlined above, is an absolute prerequisite for comprehension of the factors determining longevity of saprophytic survival by root-infecting fungi. At the outset of the work on saprophytic survival of *Ophiobolus graminis*, Botanist had anticipated that rate of its disappearance from the wheat straws would be correlated with the rate of overall microbial activity in the soil. In general terms this turned out to be true, but with respect to the effect of nitrogen the original expectation was exactly reversed. *O. graminis* was found to die out early from the rigid and little-decomposed straws buried in nitrogen-poor soils, whereas it was found to survive in almost 100% of straws in nitrogen-rich soils that had become so soft, attenuated and disintegrated by decomposition that they were difficult to handle for the wheat-seedling test. At first Botanist adopted a wrong hypothesis to explain this quite unexpected finding, but further thought upon the problem followed by microscopical examination of the straws after burial in the soil showed them that the mycelium of *O. graminis* continued to develop in the straw tissue lying within the soil, and that this development was greater, and continued for longer, in nitrogen-rich soil. This observation have been confirmed by Chambers & Flentje (1969). We can therefore conclude that *O. graminis* dies out prematurely from carbon starvation in wheat straw buried in nitrogen-poor soil, because it cannot obtained the nitrogen needed for continued mycelial growth and hydrolysis of fresh areas of cellulose within the straw tissue. The fungus is thus starving in the middle of a well-stocked larder, because it can no longer get at its food.

After this work of *O. graminis*, it was fully expected that Scientist should find a similar effect of nitrogen in prolonging saprophytic survival of other cereal foot-rot fungi, and perhaps also of other root-infecting fungi similarly surviving in other mature plant tissues with a high C/N ratio. They were therefore surprised when Butler (1953) discovered that excess of soil nitrogen actually reduced longevity of *Cochliobolus sativus* in wheat straw. Subsequently Butler (1959), continuing this work in New South Wales, confirmed this result by comparing longevity of saprophytic survival of *C. sativus* and three other cereal foot-rot fungi in a ley soil of high fertility (total N = 0.3%) and in an arable soil of low fertility (total N = 0.07%). Survival of both *O. graminis* and *Fusarium roseum* f. *cerealis* was longer in the high-nitrogen soil, whereas that of *C. sativus* was longer in the low-nitrogen soil; survival of *Curvularia ramosa* appeared to be indifferent to the level of soil fertility and nitrogen content (Fig. 4.1).

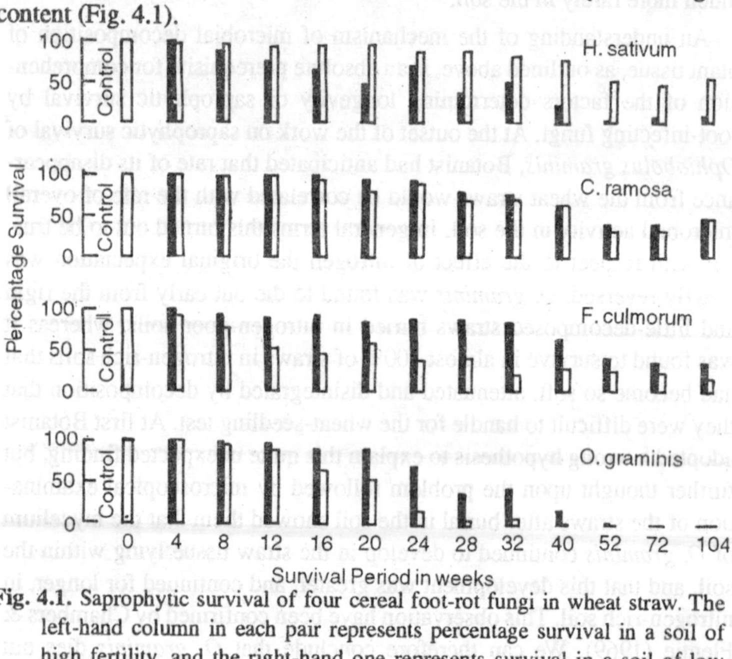


**Fig. 4.1.** Saprophytic survival of four cereal foot-rot fungi in wheat straw. The left-hand column in each pair represents percentage survival in a soil of high fertility, and the right-hand one represents survival in a soil of low fertility.

In discussing these results, Butler suggested that this effect of nitrogen in actually reducing longevity of *Cochiliobolus sativus* might be connected with its low tolerance of fungistatic growth-products excreted by competing soil micro-organisms (Butler). He supposed that this might limit the slow but continued mycelial development such as earlier it was observed for *O. graminis* in the wheat straw tissue; if further mycelial development of *C. sativus* were thus limited, then diffusion of soluble nitrogen into the straw tissues from the surrounding soil would not benefit *C. sativus*, but might well help other soil micro-organisms, and especially soil fungi, to invade the substrate, and thus to increase the population of competing micro-organisms within the straw tissue. At this stage of the discussion, it will be helpful to summarize evidence on effects of nitrogen on longevity of saprophytic survival by five cereal foot-rot fungi; in Table 4.1, a 'positive effect' of nitrogen indicates that this nutrient prolongs survival, whereas a 'negative effect' denotes the reverse.

**TABLE 4.1**

**Effects of Nitrogen on Longevity of Saprophytic Survival by Five Cereal Foot-rot Fungi**

| *Pathogen* | *Disease* | *Effect of nitrogen* | *Authors* |
|---|---|---|---|
| *Ophiobolus graminis* | Take-all | Strongly positive | Garrett (1938, 1940, 1944)<br>Butler (1953, 1959)<br>Macer (1961)<br>Chambers & Flentje (1969) |
| *Fusarium roseum* f. cerealis | Foot-rot | Strongly positive | Butler (1959) |
| *Cercosporella herpotrichoides* | Eyespot | Weakly positive | Macer (1961b) |
| *Curvularia ramosa* | Foot-rot | Weakly positive | Butler (1953) |
| | | Indifferent | Butler (1959) |
| *Cochliobolus sativus* | Foot-rot | Strongly negative | Butler (1953, 1959)<br>Garrett (1966) |

We can see in Table 4.1 that a strong positive effect of nitrogen on longevity of *O. graminis* has been repeatedly confirmed and so has a strong negative effect on longevity of *C. sativus*. The strong positive effect on *F. roseum* rests on the results of a single experiment by Butler (1959). Responses to nitrogen by *C. ramosa* and *C. herpotrichoides*, respectively, have been either weak or indifferent.

We suspected that these differing responses to nitrogen in longevity of saprophytic survival by these five fungi might be connected in some way with their respective cellulolytic abilities. The cellulose content of mature wheat straw comprises about 40% of total dry matter, yet no information on cellulolytic ability of these five species appeared to be available. To procure this, Scientists measured loss in dry weight by filter-paper cultures of the five fungi over a standard incubation period of 7 weeks at 22.5°C. *C. sativus* was found to have the highest cellulolysis rate and it seemed possible that this was somehow connected with its strong negative response to nitrogen in longevity of survival. Nevertheless, cellulolysis rate by itself could not be correlated at all clearly with the nitrogen effects; Mycologists then came to realize that the adequacy of the cellulolysis rate for continued saprophytic survival of a fungus in wheat straw tissue is not determined by this rate alone, but by the relation between this rate and the general metabolic rate of the fungus. The general metabolic rate determines the rate at which carbohydrates are consumed by the processes of respiration and growth; the cellulolysis rate determines the rate at which soluble carbohydrates are made available for the metabolism of the fungus. Only if the cellulolysis rate is *adequate* to maintain the general metabolic rate will a fungus be able to continue surviving in the straw tissue. Thus a fast-growing fungus with a high metabolic rate will require a higher rate of cellulolysis to supply its needs than will a slow-growing fungus with a low metabolic rate. To express these relationships, some scientists proposed the *Cellulolysis Adequacy Index*, which is obtained by dividing cellulolysis rate by linear growth rate (as a parameter of general metabolic rate), each expressed in suitable units. If cellulolysis rate is expressed by percentage loss in dry weight of filter-paper cultures over the standard incubation period of 7 weeks at 22.5° C. and linear growth rate as mm/24 h at 22.5° C, then the cellulolysis adequacy index is found to vary around unity for the five cereal foot-rot fungi under investigation. In Table 4.2 mean values of the index of *P. roseum*, *C. herpotrichoides* and *C. ramosa* have been taken from Garrett (1966) and the mean value for *O. graminis* from Garrett (1967); that for *C. sativus* is taken from my hitherto unpublished data (Table 4.2).

A comparison of the data given in Table 4.2 with those in Table 4.1 suggests that a low value of the cellulolysis adequacy index (hereafter referred to as the CAI) is associated with a strongly positive response to nitrogen in longevity of survival and a high value of the index with a strongly negative response. The concept of the CAI can thus explain why a fungus such as *Cochliobolus sativus* does not need extra nitrogen in order to survive in wheat straw tissue; a further extension of this hypothesis can also explain why extra nitrogen actually *shortens* its period of survival. This is because a supply of nitrogen much in excess of the level needed to keep cellulolysis going at a rate just adequate for survival is going to promote an *uneconomic* rate of cellulose consumption; this will result in a complete exhaustion of cellulose in the substrate in a shorter time than if nitrogen had been supplied at a rate just sufficient to keep the fungal colony alive and metabolizing the substrate.

**TABLE 4.2**

**Cellulolysis Adequacy Indices for Five Cereal Foot-rot Fungi**

| | *No. iso-lates tested* | *Per cent. loss in filter-paper dry wt. after 7 wks at 22.5° C* | | *Linear growth rate of fungus (mm/24 h) at 22.5° C* | *Mean value for cellulolysis adequacy index* |
|---|---|---|---|---|---|
| | | *Range* | *Mean* | | |
| *Ophiobolus graminis* | 5 | 2.1–3.6 | 2.8 | 7.3 | 0.38 |
| *Fusarium roseum* f. *cerealis* | 4 | 6.5–8.5 | 7.5 | 11.0 | 0.68 |
| *Cercosporella herpotrichoides* | 4 | 0.8–1.1 | 0.9 | 1.3 | 0.69 |
| *Curvularia ramosa* | 1 | — | 6.6 | 5.3 | 1.25 |
| *Cochiobolus sativus* | 6 | 5.4–8.4 | 7.5 | 3.6 | 2.10 |

This hypothesis thus postulates that the primary factor determining the response of a fungus to nitrogen is its own behaviour in metabolizing the cellulose of the substrate, as expressed by its CAI value, and that other fungi and other soil micro-organisms play a subsidiary role in decomposition and eventual exhaustion of the substrate. The *direct effect* of nitrogen upon the cellulolytic activity of the fungus itself thus differs from the *indirect effect* proposed by Butler (1953), who thought that nitrogen acted by assisting other micro-organisms to invade and colonize the substrate alongside its primary colonizer. These two effects are not mutually exclusive but an experiment was designed with *Cochliobolus sativus* to determine which effect was the more important. A total amount of nitrate nitrogen sufficient to curtail saprophytic survival of *C. sativus* was provided in two ways: (1) to the straws in the pure culture flasks before inoculation; (2) to the soil at time of burying the colonized straws. Scientists argued that if the effect of nitrogen was mainly on the cellulolytic activity of *C. sativus* itself (the *direct effect* hypothesis), then nitrogen would reduce longevity more if added to the pure culture flasks, because it would have a month longer in which to act on the fungus, and because most of it would be absorbed and retained by the mycelium of *C. sativus*. But if the main effect of nitrogen was to assist other fungi and micro-organisms to invade and colonize the substrate (the *indirect effect* or *replacement* hypothesis), then nitrogen would reduce longevity of *C. sativus* more if it was added to the soil at time of straw burial. The longevity of *C. sativus* was estimated on a sample of fifty straws from one soil jar for each of the six experimental treatments (two fungal isolates times three nitrogen treatments) over the period 4–16 weeks. After recovery from the soil, the encasing leaf-sheath around each straw was removed and the surface of the culm segment was well rubbed under the running tap before sand-plating. The object of this was to remove any conidia, which act as survival spores in *C. sativus*; if this precaution had not been taken, dormant survival by conidia attached to the straws might have been wrongly recorded as saprophytic survival. After 9 days incubation on the sand plates at 22.5° C, each straw was thoroughly scanned under the stereoscopic microscope (50 × magnification); any straw bearing one or more erect, sporulating conidiophores was recorded as containing still viable *C. sativus* (Table 4.3).

**TABLE 4.3**

**Effect of Nitrogen and Its Time of Application on Longevity of Saprophytic Survival by Cochliobolus Sativus in Wheat Straws Buried in Soil**

| | *Percentage straws with C. sativus still viable* | | | | |
|---|---|---|---|---|---|
| | | *Period in soil (weeks)* | | | |
| | | *4* | *8* | *12* | *16* |
| Isolate $P_1$, | no nitrogen | 100 | 100 | 92 | 82 |
| | soil nitrogen | 100 | 100 | 60 | 14 |
| | pure-culture nitrogen | 100 | 72 | 48 | 4 |
| Isolate BS790, | no nitrogen | 100 | 100 | 100 | 74 |
| | soil nitrogen | 100 | 100 | 92 | 42 |
| | pure-culture nitrogen | 100 | 100 | 78 | 2 |

The results of this experiment, displayed in Table 4.3, have returned a definite answer in favour of the *direct effect* as the more important of the two. After this, scientist realized that the most convincing evidence for the direct effect hypothesis would be to reproduce the whole spectrum of nitrogen responses in a fungus growing in pure culture upon a cellulose substrate. Although, over the range of nitrogen levels tested, a significantly negative response had been shown only by *C. sativus*, it seemed likely that the response of any of these five fungi would switch from positive to negative once the critical nitrogen level had been exceeded. The optimum nitrogen level for survival of *Ophiobolus graminis* in straw buried in soil is around 0.5 g N/100 g air-dry straw; results of an earlier experiment suggested that 0.9 g N was supra-optimal for maximum longevity of survival. *O. graminis* was selected as the subject for this experiment: (1) because its saprophytic survival is shorter than that of the other four fungi; (2) because symptoms of root infection in the wheat-seedling test are completely reliable for diagnosis. As most fungi survive in pure culture for a long time, scientist took all possible precautions to reduce the survival period in this experiment, and the substrate was reduced to a single circle of Whatman no. 3 filter paper lying on a reservoir of inorganic nutrient salts in pure quartz sand. At each sampling time, survival of *O. graminis* on the filter-paper circle was estimated by percentage infection of the roots of twelve wheat seedlings planted thereupon; percentage figures given in Table 4.4 are based on examination of some 120 roots for each treatment-series at each sampling time (Table 4.4).

**TABLE 4.4**

**Effect of nitrogen level on longevity of survival by Ophiobolus graminis in pure culture on filter-paper cellulose**

| | | *Percentage roots infected by O. graminis* | | |
|---|---|---|---|---|
| | | *Time of sampling (weeks)* | | |
| | | *10* | *15* | 23 1/2 |
| Nitrogen series | S/4 | 80 | 80 | 16 |
| | S/2 | 99 | 97 | 55 |
| | S | 98 | 99 | 92 |
| | 2S | 100 | 85 | 82 |
| | 4S | 100 | 73 | 73 |

Standard error of means = 3.79: least difference for significance at 5% level = 11.

Table 4.4 shows that the optimum level of nitrate nitrogen for maximum longevity of *O. graminis* on filter-paper cellulose was that provided by series S, which was 0.235 g N/100 g air-dry filter paper. At the final sampling after 23 1/2 weeks, the effect of the highest nitrogen treatment (0.94 g N/100 g air-dry filter paper in the 4S series) in reducing longevity of *O. graminis* was highly significant (at the 0.1% level). The results of this experiment were thus in complete accord with prediction from the *direct effect* hypothesis concerning the various effects of nitrogen on longevity of saprophytic survival.

**Other Factors Affecting Survival**

We have already mentioned the possibility that saprophytic survival of mycelium may be augmented by the production of dormant propagules in the host tissues. *Ophiobolus graminis* is not known to produce any type of dormant propagule; in the experiment with *Cochliobolus sativus* to which Table 4.3 refers, scientist had taken due precautions to separate saprophytic mycelial survival, which they were studying, from dormant survival by conidia. This precaution was necessary, because conidia of *C. sativus* functions as dormant propagules in soil; their effectiveness as such is more readily understood since the recent announcement by Meronuck & Pepper (1968) that chlamydospores are produced within the cells of conidia overwintering in soil. Since this complication of saprophytic mycelial survival by dormant survival was eliminated from the comparison between *O. graminis* and *C. sativus* in the previous section, there seems no reason to doubt that the contrasting effects of nitrogen on the longevity of these two fungi in wheat straws were indeed produced by the influence of nitrogen on saprophytic mycelial survival. But this conclusion cannot be extended without qualification to the nitrogen response of *Fusarium roseum* f. *cerealis*, the only other one of these five cereal foot-rot fungi showing a strong response (positive) to nitrogen in its longevity of survival in wheat straws (Table 4.1). Cook (1968) has observed the regular production of mycelial chlamydospores by *F. roseum* in infected wheat tissues. We therefore have to ask ourselves whether nitrogen, which increases longevity of *F. roseum* in wheat straws, does so by promoting an increased production of chlamydospores in the infected or colonized tissues. To this question we can return a cautious negative; the available evidence suggests that production of chlamydospores by *Fusarium* species, as well as that of other types of dormant propagule, is favoured by a high rather than by a low C/N ratio of plant tissues.

Although we can thus conclude that the effect of nitrogen on longevity of these cereal foot-rot fungi in wheat straw operates mainly through its influence on saprophytic mycelial behaviour, it seems most likely that overall longevity is determined by whether dormant propagules are produced and if so by their efficiency in promoting longevity. *Ophiobolus graminis* is the only one of these fungi that is not known to produce any type of dormant propagule, and it survives in wheat straw for not more than a year; Butler (1959) obtained survival of the other three fungi shown in Fig. 4.1 for a period of up to 2 years. The fifth fungus, *Cercosporella herpotrichoides*, was found by Macer (1961) to survive under some soil conditions in approx, 75% of colonized straws buried outside in the field for 3 years. This difference in longevity between *O. graminis* and *C. herpotrichoides* agrees with the experience of English farming practice; the take-all disease can usually be controlled by a one-year break in the rotation between susceptible crops, whereas the eyespot lodging disease requires a break of several years. *C. herpotrichoides* was reported by Sprague (1931) and Dickens (1964), to produce micro-sclerotia on nutrient agar, and we also have observed these bodies, but their function and efficiency as dormant propagules in wheat straw buried in soil needs to be investigated.

In view of the intensity of microbial competition for substrates in the soil, one of the most interesting facts brought to light by these and other studies on saprophytic survival is the remarkable longevity of pioneer fungal colonizers of plant tissue substrates lying in the soil. As we have just noted, *Ophiobolus graminis* survives under favourable soil conditions for up to a year in wheat straw, and its survival is not assisted by production of any dormant propagules; on the nutrient agar plate, at least, it is particularly sensitive to fungistatic growth products of other soil micro-organisms. How does this come about? Earlier in this chapter, we have emphasized the fact that when a primary parasite invades living plant tissues, host resistance to infection protects it, for a time at least, from following fungi, and so it establishes itself virtually in pure culture in the host tissues; in experiments with colonized plant tissues, this situation has been reproduced by establishing pure cultures of the inoculant fungi on sterilized host tissue. Establishment of the mycelium as a pure culture throughout the plant tissue must give the inoculant fungus a substantial initial advantage, but how is this advantage maintained when the tissue is buried in soil and open to invasion by all the surrounding soil micro-organisms? It has been suggested that production of fungistatic substances by the primary colonizer enables it to obstruct invasion of its substrate by following fungi, but this explanation could fit only a minority of the situations already investigated. Amongst the five cereal foot-rot fungi so far discussed, only *Cochliobolus sativus* is known to produce strongly fungistatic culture filtrates, to which it is itself sensitive, as demonstrated by strong staling of cultures on most nutrient agars. Such a suggestion has been advanced by Tyner (1966), on the basis of his demonstration that *C. sativus* produces a fungistatic culture filtrate when grown on ground-up wheat straw. But demonstration of such an effect during early growth of the fungus on this substrate does not tell us whether the fungus is still producing a significant concentration of fungistatic substances in the later stages of its saprophytic survival, when it is subsisting on the relatively low concentrations of sugars released by cellulolysis, and when it is in greater danger than earlier of being replaced by other fungi from the surrounding soil. Antibiotics, as they are produced in cultures of micro-organisms both experimentally and industrially, are products of 'shunt metabolism', to employ an apt description coined by Foster (1949). Antibiotics are produced by fungi in pure culture on a sufficiently rich nutrient medium when mycelial growth has been halted by some limiting factor, such as a nutrient deficiency. In this situation, the mycelium is unable to stop absorbing sugar from the culture medium, and the excess sugar is excreted as a shunt metabolite, which may have antibiotic properties, as Woodruff (1966) has explained in an admirably lucid review of antibiotic production. These conditions for antibiotic production may sometimes be satisfied when a fungus first invades a substrate of fresh plant tissue, for instance, Brian, Elson & Lowe (1956) have demonstrated production of patulin by *Penicillium expansum* Link em. Thom during its invasion of apple fruits, which have a fairly high sugar content. But it seems unlikely that the sugars released by cellulolysis of residual cellulose during the later stages of saprophytic mycelial survival in a mature plant tissue will reach a high enough concentration to permit a significant production of antibiotic. The rate of cellulolysis is self-limiting, because accumulation of sugar above a certain concentration temporarily inhibits further production of cellulases, as *adaptive* enzymes. In a study with *verticillium albo-atrum*, Talboys (1958 *c*) found that production of cellulase in the presence of cellulose was reduced by 0.1% and inhibited by 1% glucose. Similarly, Horton & Keen (1966) reported that production of cellulase by *Pyrenochaeta terrestris* (Hansen) Gorenz, Walker & Larson was repressed by 0.09% glucose.

On present evidence, therefore, we are bound to conclude that production of an antifungal antibiotic is not essential to saprophytic survival by a fungus for its full term in infected or colonized plant tissue, even through it may assist survival if conditions for its production are satisfied. Burehl, Lai & Huisman (1964) demonstrated production of an antifungal antibiotic by *Cephalosporium gramineum* in 2% cornmeal extract acidified to pH 4.8, and thought that production of this antibiotic helped *C. gramineum* to establish colonies from infected-tisue transplants on the agar medium made from this nutrient solution. Lai & Bruehl (1966) later suggested that continued production of this wide-spectrum antifungal antibiotic by *C. gramineum* in the infected wheat tissue, even at a low concentration, might prolong its saprophytic survival. But a subsequent paper by Bruehl & Lai (1966), on saprophytic survival in colonized wheat straws by four cereal foot-rot fungi and six saprophytic fungi, is entitled 'Prior-colonization as a factor in the saprophytic survival of several fungi in wheat straw'; the conclusion about this situation implied by their title coincides with my own, and with a more general one enshrined in the English proverb 'Possession is nine points of the law'.

## SURVIVAL IN TISSUES OF WOODY PERENNIAL HOSTS

A change of scene to problems of saprophytic survival by root-infecting fungi in root systems and collar regions of tree species does not necessitate an immediate change from the main topic of the preceding section, because here again the problem of nitrogen as the main mineral nutrient has attracted recent interest, as exemplified by the work of E.B. Cowling, W. Merrill and their associates in the USA. As Cowling & Merrill (1966) have pointed out, the C/N ratio of wood in different species of tree mostly varies within the range 350–500, though in some species it is as high as 1250. It is apparent, therefore, that wood-inhabiting fungi must exercise a quite remarkable economy in the deployment and conservation of so meagre a supply of nitrogen, to account for the known rapidity with which they can decompose timber. Levi, Merrill & Cowling (1968) have provided evidence for the hypothesis that these fungi conserve nitrogen by recycling it from the older and autolysing regions of the mycelium to the younger and actively metabolizing hyphae. In the presence of glucose, various fungi were found able to utilize several fractions of their own mycelia as sole source of nitrogen in pure culture. Cell-wall nitrogen was taken up less quickly than cytoplasmic nitrogen; experiments both with mycelial fractions and with model nitrogenous compounds showed that proteins, peptides and amino acids were more quickly utilized than nucleic acids, nucleotides or cell-wall fractions.

More recently, Levi & Cowling (1969) have turned their attention to the means whereby a limited supply of nitrogen may be deployed with maximum economy in the mycelia of these wood-destroying fungi. They found that when *Polyporus versicolor* L. ex Fr. was grown on media varying widely in C/N ratio, the nitrogen content of its mycelium ranged from 8.2% on a medium of C/N 4 to one of 0.1% on a medium of C/N 1600. Comparative analyses of these nitrogen-rich and nitrogen-poor mycelia showed that under conditions of nitrogen starvation, nitrogen was allocated to metabolically active cell constituents such as the nucleic acids, in preference to structural components of the cytoplasm and cell walls. A representative group of white-rot fungi was found able to produce cellulases on a cellulose medium of C/N 2000; other groups of cellulose-decomposing fungi were unable to produce cellulases at this C/N ratio of the culture medium. The white-rot group of fungi are the most highly specialized amongst wood-decomposers, inasmuch as they can degrade lignin as well as cellulose; from these data of Levi & Cowling, it seems that their specialization has extended to include adaptation to the nitrogen-poverty of their substrate.

The massive woody root systems and collar regions of tree species differ so much from the tissues of herbaceous plants as to make the problem of saprophytic survival quite different in most of its aspects. The much greater size of tree roots and stumps, as well as their mechanical construction, greatly reduces the proportion of extraneous nitrogen that can diffuse into the inner tissues from the surrounding soil, at least during the earlier stages of decomposition whilst roots and stumps are still intact. In the absence of wounds, the bark offers a mechanical barrier of cork tissue to fungal invasion; even after this barrier has been penetrated, initial invasion of the wood is limited to cellulose-composing fungi. Sylvicultural wounds that expose transverse sections of wood, however, will enable non-cellulose-decomposers, such as the blue-staining species of *Ceratocystis*, to invade the lumina of exposed vessels; another route for more rapid fungal invasion is provided by the tunnels of beetles and other wood-boring insects. Nevertheless, fungal invasion even of dead wood is slower than that of softer plant tissues; mechanical structure, chemical composition (e.g., shortage of nitrogen) and sometimes the presence of fungistatic substances, as in the heartwood of some tree species, all combine to impede the progress of fungal invasion, and to exercise a selective effect upon the species of colonizing fungi.

Most studies so far made of wood colonization by fungi have shown that considerable volumes of wood are often occupied by a single fungal colony. The situation thus often arises in dying or dead tree roots of a meeting between two fungal colonies advancing along the root in opposite directions. This situation has been experimentally studied by Rishbeth (1951) in the course of his work on *Fomes annosus*. Root lengths occupied by *F. annosus* were juxtaposed with root lengths occupied by another fungus; competing fungi thus investigated were various isolates of *Trichoderma viride (sensu* Bisby, 1939), species of *Ceratocystis* causing blue-stain of timber, and some common basidiomycete competitors of *F. annosus* found in pine stumps, including *Peniophora gigantea* (Fr.) Massee, *Stereum sanguinolentum* (Fr.) Fr. and *Polystictus abietinus* (Dickson) Fr. Amongst the pairings tested by Rishbeth, all three possible results were obtained, i.e. *F. annosus* replacing its opponent or being replaced by it, or a stalemate. Over all pairings, the competitive advantage of *F. annosus* was greater at 10° than at 15°C; this recalls a similar effect of temperature found for several cereal foot-rot fungi. Although both protagonists and conditions of competition were quite different in these two situations, they had one thing in common, i.e. that a parasite was competing with other fungi, some of which were obligate saprophytes; this effect of temperature is worth further exploration. Rishbeth also found, as might have been anticipated, that *F. annosus* was displaced less frequently from recently infected roots than from old infected roots in an advanced stage of decay. Lastly, *F. annosus* was found to survive for longer in infected sinker roots lying at some depth in the soil than in the shallower-growing lateral roots, and for longer in alkaline than in acid soils.

Replacement of one fungus by another implies the possibility of a fungal succession in such woody tissues, and such a succession of fungi in unprotected stumps was studied by Meredith (1960) in East Anglican plantations of Scots and Corsican pines. Stump surfaces exposed by tree-felling and left without any of the various protection-treatments now employed were at first infected by air-borne spores either of *F. annosus* or of one or other of two common basidiomycete competitors, *Peniophora gigantea* and *Stereum sanguinolentum*. These two latter fungi are weak parasites, inasmuch as they can invade still living stump tissues in which the defence-response to infection has been much weakened by felling of the trunk, but they are not sylvicultural pathogens like *F. annosus*. The basidiomycetes were found to be accompanied in their early invasion of the stump tissues by blue-staining species of *Ceratocystis*, which are non-cellulolytic, and thus occupy an ecological niche distinct from that of the basidiomycetes. These primary basidiomycete invaders were often found to have been replaced in 3 to 4-year-old stumps by *Hypholoma fasciculare* (Hudson) Fr., and after 5 years *Tricholoma rutilans* (Schaeffer) Fr. had become the commonest basidiomycete occupant of these stumps (Table 4.5).

**TABLE 4.5**
**Succession of basidiomycetes in pine stumps**

| | *Percentage stumps occupied by fungi listed,, at various ages from tree-felling* | | | | | |
|---|---|---|---|---|---|---|
| | *Age of stumps (years)* | | | | | |
| | *1* | *2* | *3* | *4* | *5* | *8* |
| *Fomes annosus** | 43 | 37 | 16 | 7 | 17 | 14 |
| *Peniophora gigantea** | 85 | 89 | 24 | 3 | — | — |
| *Stereum sanguinolentum** | 32 | 29 | 26 | 15 | 6 | 5 |
| *Polystictus abietinus** | — | 4 | 35 | 13 | 33 | — |
| *Hypholoma fasciculare** | — | — | 45 | 55 | 24 | 12 |
| *Tricholoma rutilans* | — | — | 15 | 43 | 73 | 66 |
| *Polyporus amorphus* | — | — | 5 | 8 | 4 | — |
| *Paxillus atrotomentosus* | — | — | — | — | 5 | 4 |
| *P. involutus* | — | — | — | — | — | 1 |
| *Collybia maculata* | — | — | — | — | 4 | 1 |
| *Clitocybe aurantiaca* | — | — | — | — | — | 3 |
| *Flammula sapinea* | — | — | — | 3 | 5 | 2 |
| *Pholiota flammans* | — | — | — | — | — | 1 |

*Species identified by mycelial characters; remaining species identified by presence of sporophores.

Despite the liability of *F. annosus* eventually to be replaced by other fungi in stump root systems, its longevity in such tissues appears to vary widely according to size of stump and soil type. On an alkaline soil, Rishbeth (1951) found *F. annosus* to be still alive in 85% of stumps of 60-year-old pines no less than 25 years after the trees had been felled.

Colonies of wood-rotting basidiomycetes are often bounded in the woody tissue by a distinctive, dark-coloured envelope. In a section across the wood, this envelope is seen as a dark-brown or black, narrow zone, and is therefore called a 'zone-line' because it is viewed in only two out of its three dimensions (and even then this term violates the geometrical definition of a line as 'having length but no breadth'). One of the earliest descriptions of zone-lines was that by Brooks (1915) of the conspicuous black zone-lines produced by *Ustulina zonata* (Lév.) Sacc. in wood of the rubber tree. Campbell (1934) has described the zone-lines formed by *Xylaria polymorpha* (Pers.) Grev. and *Armillaria mellea*. The boundary envelope of the fungal colony in the wood is formed by a tissue of dark-coloured, sclerotoid and inflated fungal cells. This boundary layer thus resembles the rind of a fungal sclerotium, and so Campbell proposed the name of *pseudosclerotium* for the whole fungal colony in the wood. It cannot be denied that Campbell's analogy is a useful one, provided that the parallel is not pushed too far and allowed to obscure the real physiological difference between these two types of fungal tissue. It is true that, by definition, a sclerotium can incorporate host tissue and even soil; nevertheless, the host tissue if present is usually ghost-like in the mature sclerotium, and its nutrients have been transferred to the fungal cells, as in sclerotia of *Helicobasidium purpureum*, formed around small lateral roots on the tuberous tap roots of sugar beet and carrot. So the mature sclerotium of *H. purpureum* and similar fungi represents *dormant survival* of an aggregation of hyphal cells, well stocked with reserves of carbon and other nutrients and comparable to a well filled larder. But in the pseudosclerotium, as defined by Campbell, the fungal larder is still in process of being filled up by active fungal metabolization of the woody host tissue; the pseudosclerotium thus seems to have more in common with saprophytic survival by a colony of still active mycelium than with dormant survival by a mature sclerotium.

To say this, however, is not to declare that the boundary envelope (or zone-line) of the pseudosclerotium lacks the protective function commonly assigned to the rind of a sclerotium; indeed, there is good evidence for the protective function of zone-lines. In a study with several wood-rotting fungi, Hopp (1938) concluded that zone-lines were formed only when the air-moisture régime in wood lay within certain limits; because air and water contents of wood are complementary, as they are in soil, water-saturated wood is oxygen-deficient, and dry wood is water-deficient. Hopp's findings were confirmed by Rishbeth (1951), who concluded that gradual desiccation promoted formation of zone-lines by *Fomes annosus* in pine roots. He found that *F. annosus* could be replaced by *Trichoderma viride* and by *Torula ligniperda* (Willkom) Sacc. in root lengths in which zone-lines had either not been formed y *F. annosus* or had been damaged; *F. annosus* was not replaced by these or other fungi when protected by intact zone-lines. Rishbeth has suggested that *F. annosus* surviving in a stump and its roots is protected from replacement by other fungi as a result of zone-lines induced by progressive desiccation, from the stump surface downwards and from the distal region of the roots upwards. Fluctuation in soil moisture content during the summer causes expansion and contraction of the wood, which is likely to result in breaching of the zone-lines, especially if wood decay due to *F. annosus* is in an advanced stage. Continuing desiccation of the decaying wood under a low enough soil moisture content then enables

*F. annosus* to form another zone-line nearer to the stump centre, and so the fungus gradually retreats in a succession of rearguard actions, marked by a successive series of zone-lines.

These conclusions about the protective function of zone-lines have been upheld by Nelson's (1964) study of survival by *Poria weirii* in naturally infected wood blocks (2 in$^3$), cut from the heartwood of a Douglas fir, and buried in soil under a young stand of the same species. A sample of thirty-six blocks was taken after 6, 9, 12 and 20 months from burial; at each sampling, presence/absence of zone-lines in each block was recorded, and viability of *P. weirii* was determined by tissue-plating. The percentage of blocks with zone-lines varied from 69 to 83% over the sampling period. *P. weirii* was recorded from only 10% blocks *without* zone-lines after 6 months in soil, and from none thereafter. But from blocks *with* zone-lines, it was recovered from 100, 90, 70 and 32% at successive sampling periods up to 20 months.

The protective function of the zone-line boundary envelope of the pseudosclerotium was an essential ingredient of the hypothesis proposed by Bliss (1951), to explain the mechanism whereby *Armillaria mellea* in infected citrus roots is killed by soil fumigation with carbon disulphide. As a background to his hypothesis, Bliss noted that *A. mellea* generally persisted for 6 years or more in infected citrus roots of not less than 3 cm diam. in Californian soils; he attributed this considerable longevity to the boundary layer of the pseudosclerotium. In the course of testing for survival of *A. mellea* in infected roots after soil fumigation, Bliss observed that *Trichoderma viride* regularly grew out from tissue platings that failed to yield *A. mellea*, and this led him to suggest that *A. mellea* had been killed not by direct fungicidal action of carbon disulphide, but by antibiotics produced by *T. viride*. Dominance of *T. viride* in soil after fumigation with quite a wide variety of fumigants had already been reported by that time. Bliss therefore suggested that this much augmented soil population of *T. viride* resulting from fumigation was able to penetrate the boundary layer of the *A. mellea* pseudosclerotium, whereas the much smaller population in unfumigated soil was unable to do so. He adduced three kinds of evidence in support of his hypothesis: (1) that the death of *A. mellea* did not immediately follow soil fumigation, at least in the stouter infected roots, but was spread out over a period of some 3 weeks; (2) that *A. mellea* in infected roots could survive a similar fumigation in sterile soil; (3) that *A. mellea* was killed if infected roots were buried in a pure culture of *T. viride*.

It was not incompatible with this very useful hypothesis to admit the possibility that carbon disulphide might kill *A. mellea* without roots of small diameter by direct fungicidal action, and that it might further damage the mycelium of this fungus in the outer part of stout roots within the limits of fumigant-diffusion, and so impair the resistance of *A. mellea* to replacement by *T. viride*. Employing woody inoculum segments of *A. melea* (approx. 1.75 cm diam.) Botanists found that *A. mellea* could be killed by direct fungicidal action of carbon disulphide in sterile soil, under which conditions no alternative explanation was possible. But Mycologist was also able to demonstrate an indirect, microbiological effect of soil fumigation, as postulated by Bliss, upon the viability of *A. mellea*. They buried similar small inoculum segments in soil that had earlier been treated with carbon disulphide at a dosage concentration and period previously ascertained to ensure a maximum development of *T. viride* in the soil after fumigation. After fumigation at this optimum dosage, the soil containers were uncovered to permit the fumigant to disperse from the soil into the atmosphere over a period of 3 weeks; during the third week, green plaques of sporulating *T. viride* could be seen developing between the soil and the glass wall of the containers. At the conclusion of this period, inoculum segments containing *A. mellea* were buried in the fumigated but now fumigant-free soil, and incubated for 23 days; the inocula were then removed and set up in glass tubes filled with soil to assay their capacity for rhizomorph production, as a test for the viability of *A. mellea* within each inoculum segment. In a comparison between twenty inoculum segments buried in unfumigated soil and twenty buried in soil fumigated with an optimum dosage of carbon disulphide, I found that thirty-nine of forty segment-ends from the control series produced rhizormorphs within 26 days in the growth tubes; in the series from the fumigated soil, only twenty-eight of forty segment-ends had produced rhizomorphs after 75 days. Incubation of the inoculum segments for 23 days in the fumigated but fumigant-free soil had thus caused 30% of segment-ends of lose their ability to produce rhizomorphs. This effect must have been due to some change in the soil brought about by fumigation; taken in conjunction with the visible development of *T. viride* at the soil-glass interfaces of the containers, it seemed to support at least some of the claims made by Bliss.

Bliss had also reported a complete kill of *A. mellea* in infected roots buried in a pure culture of *T. viride*, and this suggested exploration of the possibility that the inoculum potential of *T. viride* was a determinant of success in its lethal action against *A. mellea*, as was indeed explicitly postulated by Bliss. In an experiment to test this, scientists prepared pure-culture inoculum of *T. viride* on autoclaved soil, and aliquots of this were diluted to one-quarter and one-eighth strengths with the same soil unsterilized. Twenty inoculum segments were buried in each of these soil mixtures for 23 days, and all segments were then transferred to soil growth tubes to assay capacity of inocula to produce rhizomorphs, as before. *A. mellea* failed to produce rhizomorphs, and was therefore presumed dead, in 95% of inoculum segments buried in the pure culture of *T. viride*. Amongst the twenty inoculum segments buried in one-quarter strength inoculum of *T. viride*, 25% of segment-ends failed to produce rhizomorphs within 76 days; this loss of rhizomorph-producing capacity was thus closely comparable to that following incubation of inoculum segments in a soil that had received a fumigant dosage optimum for subsequent development of *T. viride* in the soil. Taking Bliss's results and my own together, they concluded that some post-fumigation biological control of *A. mellea* by *T. viride* could occur under optimum conditions, though less than Bliss had originally claimed. More modestly on behalf of *T. viride*, they preferred to say that *T. viride* could help to finish off what fumigation had begun. At that time some scientist did not appreciate that populations of *T. viride (sensu* Bisby, 1939) in fumigated soils could vary very widely in their antagonistic effects against *A. mellea* and other fungi, as will shortly transpire. Some scientists were lucky in obtaining positive results, though whether it is 'lucky' to obtain results that can be confirmed only with difficulty by subsequent workers is open to question.

Reasons for the development of dominance by *Trichoderma viride* (*sensu* Bisby, 1939) after soil fumigation with a suitable dosage of carbon disulphide have been explored by Saksena (1960) and Moubasher (1963). Both these workers found in fumigation trials with a range of carbon disulphide doses that *T. viride* possesses only a moderate degree of tolerance to carbon disulphide; various species of *Penicillium* and *Aspergillus*, mostly ascosporis, can survive a dose at least five times that tolerated by *T. viride*. Saksena therefore concluded that *T. viride* owes its ability to dominate the fungal population recolonizing fumigated soil to possession of a moderate but sufficient degree of fumigant tolerance allied with a high growth rate through fumigated soil; some fungi with a still higher growth rate possess only a low fumigant tolerance whereas the highly tolerant fungi are mostly rather slow growers. This combination of these two characteristics thus possessed by *T. viride* appeared to give it a unique advantage for recolonization of fumigated soil, at least amongst the common soil fungi studied by Saksena and Moubasher. The fuller data collected by Moubasher are compatible with Saksena's explanation, and he also provided a useful quantitative expression for *recolonizing ability*, in the ratio obtained by dividing the soil population of a particular fungal species 36 days after fumigation by its population 1 day after fumigation. He estimated soil populations by soil-dilution-plate counts. This recolonizing ratio for *T. viride* was 9.2, and was higher than that for any of the other twenty-three common soil fungi that Moubasher investigated.

Analysis of this problem was carried further by Mughogho (1968), who began work in the hope that he could find a fumigant more reliable for promoting dominance of *T. viride* than we had found carbon disulphide to be. To obtain even a moderate development (as compared with that in sterile soil) of *T. viride* in fumigated soil, the dosage of carbon disulphide had to be kept within a narrow range and the prospects of being able to do this by soil fumigation in the field were not encouraging. Moreover, in a fumigation experiments with carbon disulphide, the indirect component of the fumigation effect on *A. mellea*, although significant, had not been good enough for a control measure in actual practice, and scientists hoped to improve on this by finding another fumigant more effective for this particular purpose. Mughogho's first results in this search were encouraging, as the following fumigants were found superior to carbon disulphide for promoting dominance of *T. viride*, both in respect of breadth of suitable dosage range, and in the size of *T. viride* populations developing in the fumigated soils: allyl alcohol, chloropicrin, methyl bromide, D-D mixture, sorbic acid and acetylenedicarboxylic acid. It was therefore unexpectedly disappointing for Mughogho to find that these high populations of *T. viride* in the fumigated soils more often than not failed to exert any controlling effect either on production of rhizomorphs by *Armillaria mellea* or on mycelial growth through soil and infection of swede seedlings by *Rhizoctonia solani*. Most fortunately, however, it became possible to explain these unexpected results through the current research by J. Webster and M.A. Rifai at the University of Sheffield on a revision of the taxonomy of the genus *Trichoderma*. In 1939, G.R. Bisby revised the taxonomy of the genus *Trichoderma* Pers. ex. Fr. and incorporated all described species into *Trichoderma* Pers. ex Fr., thus making the genus a monotypic one. But Webster and Rifai have now found good taxonomic reasons for putting asunder what Bisby had joined together, and Rifai (1969) has published a conspectus of the genus *Trichoderma* in which nine species are recognized and keyed for identification.

The species, as now recognized by Rifai (1969), that Mughogho found to make up his populations of *Trichoderma* in these variously treated soils were *T. hamatum* (Bon.) Bain., *T. harzianum* Rifai, *T. koningii* Oud., and *T. viride* Pers. ex S.F. Gray; amongst these four species, *T. harzianum* usually predominated. He further found that different isolates of *Trichoderma*, even within a single species, varied widely in their degree of antagonism towards *Armillaria mellea* and *Rhizoctonia solani*. He concluded that dominant populations of *Trichoderma* in his fumigated soils must have been made up largely of isolates that were not effectively antagonistic towards either *A. mellea* or *R. solani*. The highest proportion of effective antagonists amongst isolates of *Trichoderma* tested by Mughogho belonged to *T. viride* Pers. ex. S.F. Gray, as now delimited by Rifai (1969).

It cannot be denied that all these studies of the taxonomy and ecology of *Trichoderma* species in soil have greatly increased our comprehension of the sequence of mycological events occurring after soil fumigation; equally, however, we have to admit that our hopes of mediating biological control of root-disease pathogens by *Trichoderma* species through soil fumigation, originally aroused by D.E. Bliss in 1951, have received a set-back.

# 5

# FUNGI AS DECOMPOSERS OF LEAVES

Leaves of all manner of types form suitable substrates for many fungi. Long before any leaf falls, the complex process of its decay is initiated. The greater part of this process takes place above ground in the litter covering the soil surface. In this chapter some facets of this are discussed but particularly the epiphytic leaf surface or phylloplane micro-flora, the leaf surface as a habitat niche for fungi, colonization by the common primary saprotrophs as leaves senesce and die, the attributes which make these primary saprotrophs such widespread and successful colonizers, and some of the subsequent events occurring in the litter.

## THE LEAF AS A SPORE TRAP

As any leaf unfolds it is a relatively clean sheet which immediately provides landing sites for air-borne particles such as bacteria, yeast cells and fungal spores but also pollen. Spore trapping by leaves is a natural phenomenon of nature. Spores may reach leaves in three main ways: wind-borne and deposited by impaction or by sedimentation under gravity; in falling rain drops; or in rain splash droplets. Air-borne spores are usually dry and often rough or spiny and readily detachable from their stalks, excellent examples being the urediospores of rust fungi. They are readily washed out of air by falling rain drops. Rain splashed spores tend to be wet or slimy and borne in a sticky liquid. Adaptations facilitating deposition are far less obvious than in spores of aquatic fungi. Amongst dry spores, a larger size, as seen in the powdery and downy mildews, favours impaction and sedimentation. Rain splashed spores tend to be smaller and spherical. But these are only generalizations and there are many anomalies. The most ubiquitous and by far the most numerous of the phylloplane fungi are members of the Sporobolomycetaceae, the shadow yeasts. They produce air-borne spores which in relative terms are quite minute.

Leaf surfaces are differential spore traps. Their efficiency as traps depends upon whether they are horizontal or vertical, wet or dry, hairy or glabrous, glossy or mat, waxy or non-waxy and so on. Not all spores that land become securely attached. Some are washed off by rain, blown off by wind or redistributed by dew. Some have a two phase dispersal system. For example, the large sporangia of pathogenic species of *Phytophthora* and some other Oomycete Peronosporales are wind-borne and normally impacted onto leaf surfaces. Under moist conditions the impacted sporangia may germinate directly by a germ tube or indirectly to produce motile zoospores which may swim about in moisture or be redispersed further to other leaves in rain splash droplets.

Virtually any spore which may become air-borne can be found on leaves. If leaf surfaces are washed and the washings plated out onto nutrient agar, numerous yeasts and filamentous conidial Ascomycotina, some Zygomycete Mucorales and the occasional Mastigomycotina and Basidiomycotina develop in the plates. But microscopic examination of stained leaf surface impressions or peels reveals the presence of not only these but also spores of many other Ascomycotina and Basidiomycotina, including those of agarics, polypores and Gasteromycetes. They just do not grow or grow too slowly on the culture medium used. These impressions of peels can be made by spraying leaves with cellulose acetate in amyl acetate or painting with nail varnish or molten 1% agar, leaving to dry and then stripping off.

Many of these fungi and an even larger number of bacteria actively grow on the surface of the living leaf and have been called 'resident inhabitants' in contrast to 'casual inhabitants' which are unable to grow in such an environment because of the lack of essential nutrients, unfavourable physical factors, competition with or antagonism by others or some combination of these factors. This rather simple distinction can be extended by dividing the epiphytic fungi into three categories: non-pathogenic epiphytes; pathogens; and exochthonous or casual inhabitants. Exochthonous is a fitting, if somewhat clumsy, terms as it is used for fungi found on or in a substrate which is not their habitual one.

## PHYLLOPLANE INHABITANTS

Amongst the non-pathogenic epiphytes, two main groups, the phylloplane inhabitants and the common primary saprotrophs, can be recognized. The phylloplane inhabitants are able to complete their life cycle or a significant part of it on the living leaf without damaging it. *Sporobolomyces roseus* (Fig. 5.1a) not only is a very good example of such a fungus but is virtually omnipresent, being found on leaves of grasses, dicotyledonous herbs, trees and shrubs, wherever they grow. Its cells multiply very quickly by budding when conditions are favourable, forming distinct yeast-like colonies on the leaves. Budded cells can be redistributed on an individual leaf or from leaf to leaf by rain splash and are similarly locally dispersed to leaves of other plants. It also reproduces by ballistospore formation. The ballistospores are very effectively wind dispersed. They are produced under high humidities at night, as are the budded cells, and constitute the major component of the air-spora at that time. Other members of the Sporobolomycetaceae are also common. Species of *Bullera* behave similarly, whereas members of the genera *Tilletiopsis* and *Itersonilia* produce a sparse mycelium from which ballistospores arise (Fig. 5.1b). Other yeasts, especially members of the Cryptococcaceae, the non-sporing yeasts, exist in the budding phase. They all complete their life cycle in the phylloplane. Such fungi are often called 'shadow yeasts'. Their presence can be demonstrated by suspending leaves from the underside of a Petri dish lid for about 12 h over 2% malt extract agar. In such a humid and still atmosphere, ballistospores are produced and discharged. They fall vertically onto the agar below and start budding. Tiny colonies, pink in *S. roseus*, become visible after 2–3 days. These form a mirror image of the distribution of the cells of the leaf.

Two conidial Ascomycotina also grow in the phylloplane. The conidia of *Aureobasidium pullulans* and several species of *Cladosporium* may germinate after impaction and develop into hyphae forming quite extensive colonies under favourable conditions. *Aureobasidium* more often grows by yeast-like budding with minimal hyphal growth (Fig. 5.2a). This is a modification of its normal cultural form. The budded cells may again be redistributed in moisture films or by rain splash droplets. The conidia of *Cladosporium* may germinate and produce secondary conidia from short germ-tubes rather than grows as hyphae (Fig. 5.2c). These conidia are dry and become air-borne. So both reproduce rapidly and complete a significant part of their life cycle in the phylloplane. Both

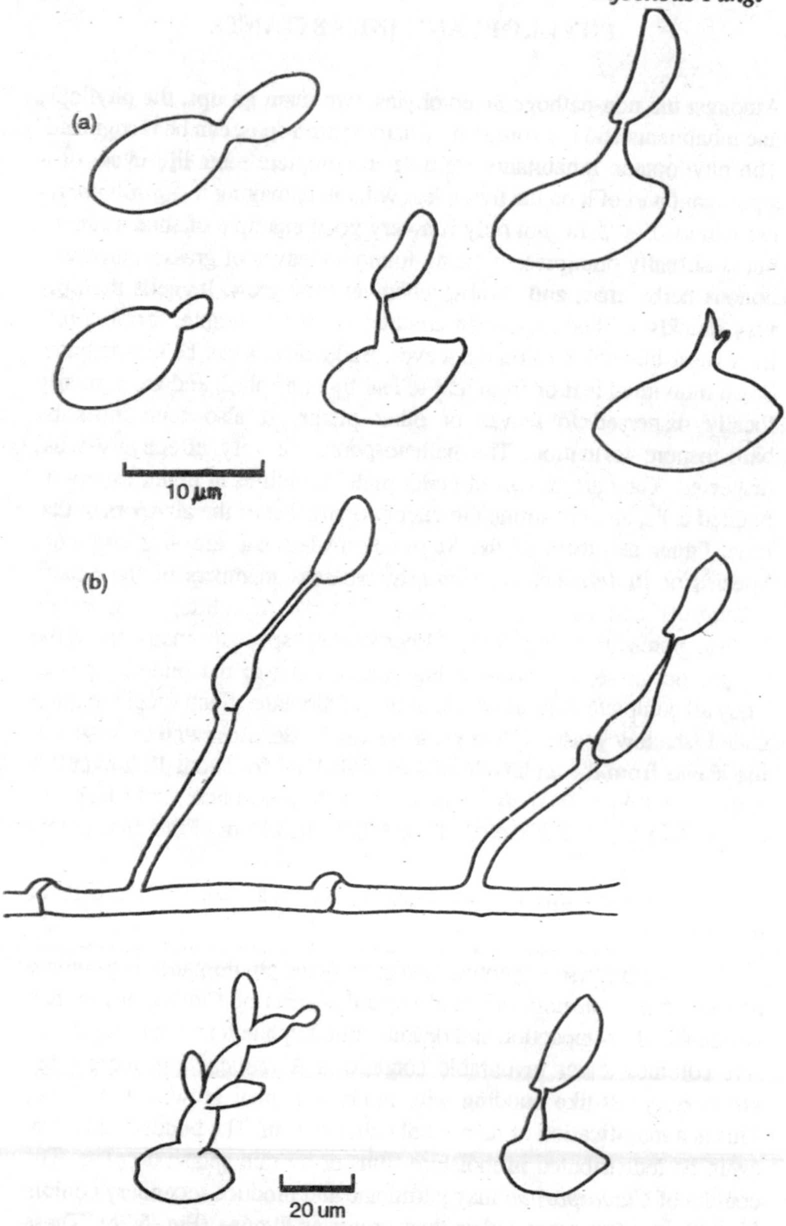


**Fig. 5.1.** (*a*) **Budding and ballistospore formation in *Sporobolomyces roseus*.** (*b*) **Ballistospore formation in *Itersonilia perplexans*, ballistospores germinating by budding and ballistospore formation.**

eventually produce ascocarps to complete their life cycle. These are found only in the spring on overwintered fallen leaves in temperature climates. Ascospores are discharged from these as new leaves unfold.

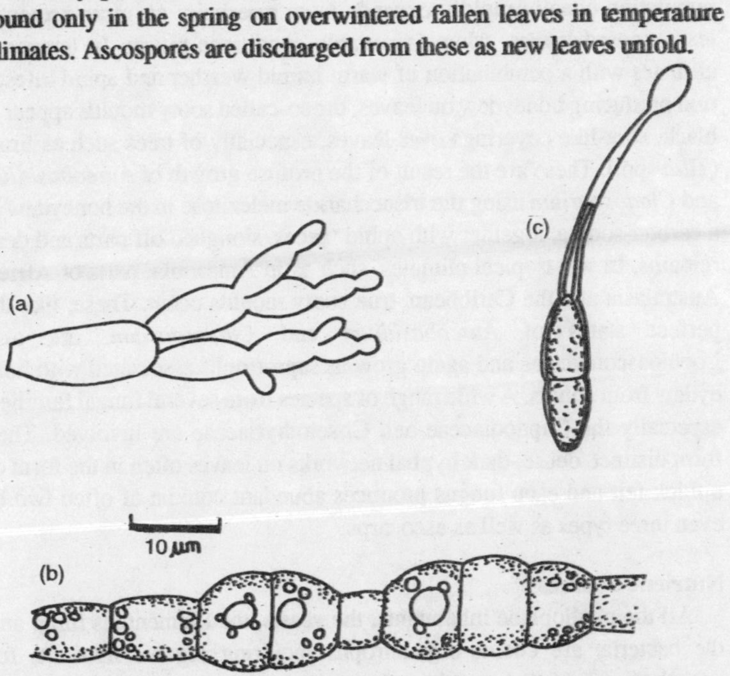


**Fig. 5.2.** (*a*) Yeast-like budding by *Aureobasidium; and* (*b*) chlamydospores of *Aureobasidium* on a leaf furface. (*c*) Conidium of *Cladosporium* germinating to produce a secondary conidium.

*Aureobasidium* and *Cladosporium* are also both very well-adapted to survive in this rigorous habitat. Their hyphal walls rapidly become thickened and melanized. This enables them not only to survive exposure to damaging ultra-violet light from the sun but may also help to prevent excessive desiccation and make them more resistant to bacterial lysis. *Aureobasidium* produces dark, thick-walled multicellular chlamydospores in chains or in clumps (Fig. 5.2b). *Cladosporium* produces more distinct microsclerotia, compact spheres of 10–100 cells with an outer layer of thick-walled cells with heavily melanized walls. Under favourable conditions, these produce clusters of conidiophores and abundant conidia which, like the ascospores, serve as a source of inoculum as new leaves unfold. In contrast *Sporobolomyces* does not appear to be able to withstand prolonged adverse conditions, such as low relative humidities. At relative humidities of 65% and below, it rapidly disappears but its population equally rapidly expands from reservoirs on more protected less exposed leaves, when favourable conditions return. In temperate climates with a combination of warm humid weather and aphid infestation producing honeydew on leaves, the so-called sooty moulds appear as black, soot-like coverings over leaves, especially of trees such as limes (*Tilia* spp.). These are the result of the profuse growth of *Aureobasidium* and *Cladosporium* using the trisaccharide melezitose in the honeydew as a carbon-source, together with aphid faeces, sloughed off parts and dead remains. In wet tropical climates, such as in Amazonia, parts of Africa, Australasia and the Caribbean, true sooty moulds occur. These, like the perfect states of *Aureobasidium* and *Cladosporium*, are also Loculoascomycetes and again grow as saprotrophs associated with honeydew from aphids. A wide range of species from several fungal families, especially the Capnodiaceae and Chaetothyriaceae are involved. They form distinct, dense, dark hyphal networks on leaves often in the form of a thick felt and each fungus produces abundant conidia of often two or even three types as well as ascocarps.

**Nutrient Sources**

All the phylloplane inhabitants, the yeasts, the filamentous fungi and the bacteria, are chemo-organotrophs requiring organic nutrients for growth. Some of their nutritional requirements may be met by organic substances absorbed or deposited onto leaves such as detritus trapped in their superficial wefts of hyphae, as also happens with fungi living on paint films or glass. Most of their nutrients, however, must be derived directly or perhaps indirectly from the host. A great variety of substances exude or leak out of leaves. These include free sugars, amino acids and inorganic ions which are all essential for fungal and bacterial growth. For example, water droplets placed on leaves exhibit an increase in conductivity indicating exudation from the leaf; increased growth of some fungi in these drops shows that certain of these exudates are of nutritional value. Two sources of added nutrients are from pollen and other spores. Nutrients also leak out of these. Pollen added to leaf surfaces stimulates the development of *Sporobolomyces* and *Cladosporium* and probably accounts for the sudden increase in their population shortly after flowering on leaves of plants such as rye. Conidia of *Botrytis cinerea* placed in droplets on leaves leak out amino acids and sugars in sufficient quantity for phylloplane bacteria to develop in such numbers as to inhibit germination of the conidia themselves. This all occurs on the intact surfaces of healthy leaves. On aphid infested leaves nutrients may be derived indirectly from the host. Host sucrose is converted to melezitose in honey-dew and this is used by the phylloplane inhabitants.

The number of phylloplane inhabitants increases with the age of the leaf. This association between population density and age of the leaf is usually explained by increase in leaf exudates with ageing. It is also assumed that the restricted availability of the nutrients is one of the main causes of the relatively poor development of the phylloplane inhabitants on immature leaves. There is also evidence that some of these fungi can slowly degrade the surface waxes and cuticle and so gradually increase the permeability of the epidermis. Their numbers are also far greater on leaves infected by pathogenic fungi, such as rusts and mildews. Four to five times as many colonies of *Sporobolomyces* can be isolated from mint leaves infected with the rust fungus, *Puccinia menthae*, as from healthy mint leaves. Here the injurious effect of the pathogen, especially perhaps the changes in cell permeability, cause an outflow of additional nutrients.

The phylloplane inhabitants are also not uniformly distributed over the leaf surface. Most are more prevalent on the upper surface and are usually more predominant along the veins, frequently with their cells orientated to lie parallel with the vein axis. They also tend to align themselves along the anticlinal walls, as with veins there is a slight depression there. They could be washed into these positions but there may also be more exudates released along the veins; also vein sheath cells may bring nutrients nearer the surface and thus facilitate exudation.

## COMMON PRIMARY SAPROTROPHS

The common primary saprotrophs are unable to grow to their full extent in the phylloplane until the onset of senescence. Their pattern of development is restricted until senescence and several rarely or never grow on the green leaf. Their spores accumulate on the leaf prior to senescence and remain dormant until the death of the tissues. If they do germinate they do so only to a limited extent. On senescence they very quickly take advantage of the changing conditions. Sporing colonies of these fungi are ubiquitous on newly dead leaves of the majority of plants. The phylloplane inhabitants and the common primary saprotrophs by no means form distinct groups. *Aureobasidium* and *Cladosporium* have to be included in both groups because, although they grow and reproduce by conidia in the phylloplane, they develop to a much greater extent in the dead leaf. Other fungi, all conidial Ascomycotina, in this group include *Alternaria alternata, Botrytis cinerea, Epicoccum purpurascens* and *Stemphylium botryosum*. In the Tropics the list can be extended to include species of *Curvularia* and *Nigrospora*. Spores of a great variety of other saprotrophs may also be present on the leaves. They germinate only on the death of the leaves or sometimes thereafter.

## PATHOGENS

Two distinct categories can be recognized amongst the pathogens found on leaves. There are those from the Plectomycete Erysiphates, the powdery mildews, which are wholly restricted to the phylloplane except for haustoria in the epidermal cells of the host leaf. All their very extensive mycelium, conidia and ascocarps are borne on the leaf surface. The second category, covering virtually all other pathogens, infect leaves and grow almost entirely within them with only their reproductive structures having access to or being produced on the outside. These latter exhibit all gradations, from those which produce an appressorium from their spore and penetrate immediately, to those which have a prolonged and relatively extensive phase of epiphytic non-parasitic hyphal growth on the leaf surface before they penetrate (Fig. 5.3). The spores of many of these pathogens remain dormant for considerable periods only germinating as host resistance begins to fall prior to senescence or following a suitable change in the weather. Again no clear distinction can be drawn between pathogens of this latter type and some of the common primary saprotrophs. *Botrytis cinerea* is one of the latter but is also a necrotrophic parasite of some hosts under particular environmental conditions, such as prolonged very high humidities which favour it but not its host. Its conidia may then germinate on the leaf surface and after a phase of epiphytic growth penetrate and bring about a soft watery rot by means of its pectolytic enzymes.

## EXOCHTHONOUS FUNGI

Spores of pathogens which are unable to infect the leaves on which they have landed may also be present. They may remain dormant or they may germinate before they recognize that they are on the wrong host. They may contribute, like pollen, to the nutrients available on the leaf surface. They could be included with the exochthonous or casual fungi as they are found on leaves but do not grow there. The latter are unable to gain any nutritional advantage from the habitat which is clearly a dead end for many but by no means all. Any soil fungi with air-borne spores may

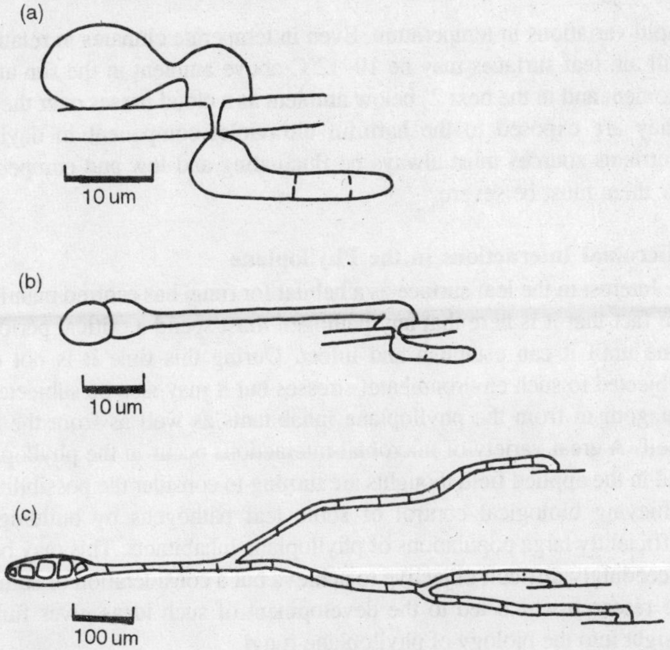


**Fig. 5.3.** Three different types of growth shown by pathogenic leaf-inhabiting fungi on leaf surfaces. (*a*) *Botrytis fabae*. (*b*) *Mycosphaerella ligulicola*. (*c*) *Cochliobolus sativus*.

be trapped on leaves and later washed off by rain onto the soil beneath and so they are successfully dispersed. Spores of many coprophilous fungi on herbivore dung are discharged onto grass leaves surrounding the dung and remain there until the grass is eaten by herbivores. Passage through the gut of a herbivore may be necessary to trigger-off their germination. Direct dung to dung dispersal is abortive. Thus impaction onto leaf surfaces is important if they are to complete their dispersal and life cycle.

## THE LEAF SURFACE AS A HABITAT FOR FUNGI

The leaf surface is a most inhospitable niche in both physical and chemical terms for fungi. Although transpiration may mitigate against extreme low levels of relative humidity, the fungi are repeatedly dried by the sun and wind and re-wetted by rain and dew. They are not insulated against temperature fluctuations and as such are subjected to marked and very rapid variations in temperature. Even in temperate climates in relatively still air, leaf surfaces may be 10–12°C above ambient in the sun at one moment and in the next 2° below ambient as a cloud passes over the sun. They are exposed to the harmful ultraviolet component in daylight. Nutrients sources must always be fluctuating and low and competition for them must be severe.

**Microbial Interactions in the Phylloplane**

Interest in the leaf surface as a habitat for fungi has centred mainly on the fact that it is here that any pathogen must spend a critical period of time until it can establish and infect. During this time it is not only subjected to such environmental stresses but it may also be subjected to antagonism from the phylloplane inhabitants as well as from the host itself. A great variety of microbial interactions occur in the phylloplane and in the applied field thoughts are turning to consider the possibility of achieving biological control of some leaf pathogens by building up sufficiently large populations of phylloplane inhabitants. This may be an exceedingly difficult objective to achieve but a consideration of some of the research which led to the development of such ideas gives further insight into the biology of phylloplane fungi.

*Pollen as a Nutrient Source and Competition for Nutrients.* The fact that phylloplane inhabitants such as *Cladosporium* and *Sporobolomyces* could benefit from nutrients leaked from pollen grains was convincingly demonstrated by Fokkema (1971). Rye leaves from two separate plots, in one of which the plants had their inflorescences removed or covered, so that no pollen fell onto the leaves below, were taken from plants and washed, twice a week from early June to September, in 1968 and 1969. The washings were plated out onto nutrient agar and the colonies of *Cladosporium* spp. which developed were counted, the assumption being made that each colony arose from a single spore (Fig. 5.4). In 1969, the number of colonies from leaves with pollen rose from 15 to 13,000 cm$^{-2}$ two weeks after flowering. On leaves without pollen the numbers were 10 and 550 respectively. On leaf senescence the colonies recorded from all leaves reached the same levels. The leaves at this stage leaked more nutrients and the stimulating effect of the pollen wore off. The larger number of colonies of *Cladosporium* recorded in 1969 was shown to be due to more pollen on the leaves, a mean of 3450 cm$^{-2}$ as against 300 cm$^{-2}$. There was frequent rain after flowering in 1968 so that the pollen was washed off. Stimulation in the presence of pollen was

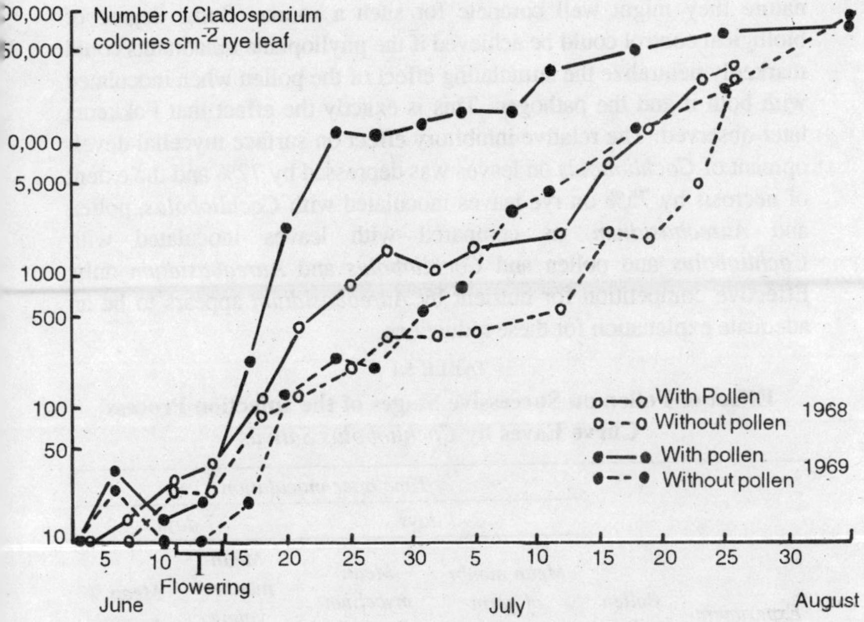


**Fig. 5.4.** The successive changes in the number of *Cladosporium* spp. colonies per cm$^2$ rye leaf during the season. The data are the means of the numbers of colonies from washings of eight penultimate leaves.

not restricted to *Cladosporium*. *Aureobasidium pullulans* and *Sporobolomyces roseus* were also stimulated. For the latter, two weeks after flowering, 33,600 colonies cm$^{-2}$ were recorded from rye leaves with pollen and only 3800 colonies cm$^{-2}$ from rye leaves without pollen.

The Loculoascomycete *Cochliobolus sativus* is a leaf pathogen of rye which makes a variable amount of epiphytic growth before penetration into the leaf (Fig. 5.3). Fokkema inoculated rye leaves with conidia of *Cochliobolus*, together with pollen and without pollen. The effect of the pollen on successive stages of the infection process and the resultant necrosis is given in Table 5.1. Leaves inoculated with *Cochliobolus* and pollen had a significantly larger percentage of necrotic areas. He attributed this to the pollen leaking out nutrients which greatly stimulated the superficial growth of the mycelium. This increased the inoculum and so more young lesions per unit area were obtained and hence eventual necrosis. In these experiments both the phylloplane inhabitants and the pathogen were relying, in part at least, on the same nutrient source and in nature they might well compete for such a source. Some degree of biological control could be achieved if the phylloplane inhabitants could markedly neutralize the stimulating effect of the pollen when inoculated with both it and the pathogen. This is exactly the effect that Fokkema later observed. The relative inhibitory effect on surface mycelial development of *Cochliobolus* on leaves was depressed by 72% and the extent of necrosis by 75% on rye leaves inoculated with *Cochliobolus*, pollen and *Aureobasidium*, as compared with leaves inoculated with *Cochliobolus* and pollen and *Cochliobolus* and *Aureobasidium* only. Effective competition for nutrient by *Aureobasidium* appears to be an adequate explanation for these reductions.

**TABLE 5.1**

**Effect of Pollen on Successive Stages of the Infection Process Curve Eaves by *Cochliobolus Sativus***

| | | *Time after inoculation* | | | |
|---|---|---|---|---|---|
| | | *2–3 days* | | *7 days* | |
| *Experiment* | *Pollen addition* | *Mean numbr of germ tubes/100 spores* | *Mean mycelium length in* μm $mm^{-2}$ | *Mean number young lesions* 10 $cm^{-2}$ | *Mean % necrotic area* |
| 1 | – | 23 | 0 | 17 | 2 |
| | + | 137 | 3600 | 113 | 58 |
| 2 | – | 67 | 250 | 22 | 2 |
| | + | 102 | 3250 | 56 | 35 |

*Antagonistic Reactions*. A number of such interactions have been reported but competition for nutrients is not the sole explanation for some of these. In some cases inhibitory substances produced by the phylloplane inhabitants may also be involved and normal leaf exudates rather than pollen may be the nutrient source. Pace and Campbell (1974) found that *Aureobasidium pullulans* and *Epicoccum purpurascens* were common in the phylloplane of *Brassica* spp. and that they were antagonistic to the wound parasite *Alternaria brassicicola* in culture. Their growing colonies inhibited mycelial growth of *A. brassicicola* and their germinating conidia inhibited the germination of its conidia. They inoculated leaves of cabbage and Brussels sprout after wounding with *Alternaria brassicicola, Aureobasidium pullulans* and *Epicoccum purpurascens* separately and with A. brassicicola plus *A. pullulans* and *A. brassicicola* plus *E. purpurascens*. There was an 80–100% infection with *A. brassicicola* alone but none with the two saprotrophs alone. The percentage of successful infections by *A. brassicicola* was reduced when it was inoculated with either of the two saprotrophs (Fig. 5.5). The

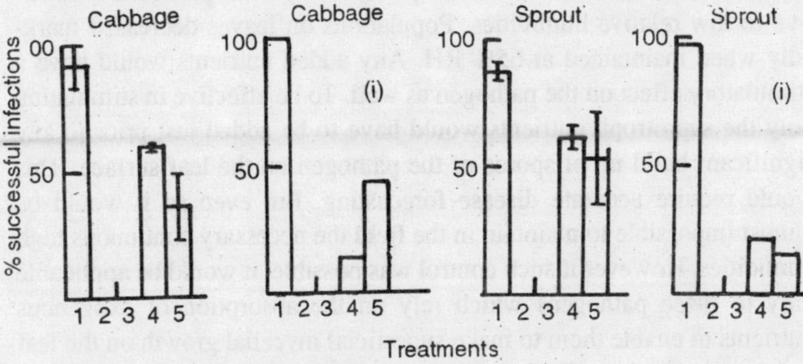


**Fig. 5.5.** Percentage (mean of three values) of successful infectioons obtained on cabbage nad Brussels sprout leaves. Treatments: 1 = *Alternaria brassicicola* alone; 2 = *Epicoccum purpurascens* alone; 3 = *Aureobasidium pullulans* alone; 4 = *A. brassicicola* + *E. purpurascens*; 5 = *A. brassicicola* + Au.pullulans. The vertical bars are the standard deviations from the mean. (*i*) = after pre-incubating the antagonist for 14h.

reduction was greater when the saprotrophs were inoculated 14 h prior to *A. brassicicola*. Both *A. pullulans* and *E. purpurascens* were capable of active growth on the leaf surface so might compete with the pathogen for nutrients in the form of leaf exudates. This could explain why they were more effective when inoculated 14 h prior to the pathogen. But they may also produce inhibitory substances. Some evidence for this is that a 50% reduction in successful infections were obtained when conidia of *A. brassicicola* were suspended in a culture filtrate of *A. pullulans* and used instead of water for inoculation. The culture medium itself enhanced infection.

**Towards biological control**

These sorts of antagonism must obviously be having some effect in the field but the question to be answered in terms of achieving biological control is how can their effects be maximized? One approach is to manipulate the system to stimulate the phylloplane inhabitants. Biological control using the indigenous population, rather than introducing others, would be possible if a sufficiently large population of antagonistic phylloplane inhabitants could be built up. Bashi and Fokkema (1977) have shown that continuous high humidities and nutrients in excess of those exuded by leaves are necessary to maintain a phylloplane population of *Sporobolomyces* dense enough to have sufficient antagonistic potential to control *Cochliobolus. Sporobolomyces* is particularly sensitive to low relative humidities. Populations on leaves decreased markedly when maintained at 65% RH. Any added nutrients would have a stimulatory effect on the pathogen as well. To be effective in stimulating only the saprotroph nutrients would have to be added just prior to any significant build up of spores of the pathogen on the leaf surface. This would require accurate disease forecasting. But even so it would be almost impossible to maintain in the field the necessary continuous high humidities. However if such control was possible, it would be applicable only to those pathogens which rely on the absorption of exogenous nutrients to enable them to make superficial mycelial growth on the leaf before penetration. Pathogens which normally penetrate the leaf immediately after germination or which have a very restricted superficial mycelial growth are probably less susceptible to competition for nutrients.

The alternative approach is to allow the natural control measures to proceed and to avoid the indiscriminate use of fungicides which may affect the phylloplane inhabitants more than the pathogen. Pace and Campbell (1974) noted that the systemic fungicide Benomyl gave a good control of many diseases but not the leaf spot of brassicas caused by *Alternaria brassicicola*. The pathogen is resistant to it. The two antagonistic saprotrophs which they used, *Aureobasidium pullulans* and *Epicoccum purpurascens* are inhibited by Benomyl. Therefore use of this fungicide could make the disease worse. Fokkema found that *Cochliobolus* is also relatively resistant to Benomyl, and that inoculation of rye leaves with *Cochliobolus* just after flowering (i.e., with pollen), resulted in 60% less necrosis on water-sprayed leaves than on Benomyl-sprayed leaves. At the time water-sprayed leaves had a natural phylloplane population of 10,000 spores $cm^{-2}$ and Benomyl-sprayed leaves only 1200 spores $cm^{-2}$. This implies that the Benomyl had reduced the antagonistic capacity of the phylloplane inhabitants but it also provides direct field evidence for naturally occurring biological control.

*Antagonism via Lysis, Antibiotic production or pH changes.* Other forms of antagonism are exhibited in the phylloplane. Bacteria may lyse fungal spores. Chitinolytic enzymes are usually involved. Lenne and

Parberry (1976) noted clusters of bacteria surrounding lysed conidia and germ tubes of the pathogen *Colletotrichum gloeosporioides* on leaf surfaces. Appressoria are necessary for penetration to occur. The bacteria failed to lyse these. They have melanized walls and there are numerous reports of melanized structures resisting the lytic action of bacteria. The production of appressoria was enhanced in the presence of bacteria but was reduced by added nutrients, such as 1% glucose peptone solution. The stimulated production of appressoria in the presence of bacteria is a normal response of the fungus to a hostile environment. Desiccation and starvation also cause appressorial formation. This response serves as an important short term survival role during the infection phase. It should be noted that in this particular case added nutrients increased germ tube growth but fewer appressoria were formed. Since the latter are necessary for penetration, added nutrients may in this case enhance disease control.

Several phylloplane inhabitants, such as *Aureobasidium* and *Sporobolomyces* have been shown to produce antibiotics in culture although there is no direct evidence that they play a role *in vivo*. Although antibiotic production by bacteria on leaf surfaces does not appear to be very widerspread, some bacteria have been shown to produce antifungal peptides which, under experimental conditions at least, reduce incidence of disease caused by a number of species of *Colletotrichum*.

Some fungal leaf pathogens are very sensitive to pH changes. In *Septoria nodorum*, for instance, spore germination is inhibited below pH 6. Conidia of this fungus, placed around the edge of a growing colony of *Botrytis cinerea*, failed to germinate. The pH of the medium fell to below 6 in advance of the hyphal tips of *Botrytis*. Such a mechanism could operate in the phylloplane.

**Fungistatic substances produced by leaves**

In addition to nutrients, leaves of many plants may exude fungistatic substances which cause inhibition of spore germination or restriction of germ tube growth. Phenols are the most widely known fungistatic substances produced by leaves. They are responsible for the inhibition of spore germination of the apple scab fungus, *Venturia inaequalis*, on some apple cultivars. Gallic acid has been identified as an antifungal component in droplets of dew obtained from sycamore leaves. Apart from these substances formed within the leaf cells and exuded onto the surface, some constituents of the cuticular waxes may also be fungistatic. An acidic ether-soluble fraction from the wax of apple leaves inhibits the growth of the apple mildew fungus, *Podosphera leucotricha*. The properties of waxes on leaves will also affect the exudation of both nutrients and anti-fungal substances. Waxes with high factor of more hydrophobic constituents will tend to limit the movement of exudates to the surface.

Thus at the leaf surface a series of complex interactions occur between pathogen/host/phylloplane inhabitants/environment. Numerous aspects and the outcome of many of these interactions are still to be discovered, but it is evident that the phylloplane inhabitants act in some sort of buffering capacity against some pathogenic fungi at least.

## DISTRIBUTION OF THE COMMON PRIMARY SAPROTROPHS

Eventually the leaf senesces either naturally or prematurely after supporting, in some cases, one or more pathogens. Of the multitude of fungal spores of a vast array of a species which are impacted onto leaf surfaces only relatively few succeed in colonizing the leaves as they senesce and grow as active saprotrophs within the leaf tissues after death. These common primary saprotrophs are virtually ubiquitous colonizers. On most leaves such as those of deciduous trees, shrubs, herbs, grasses including cereals and even bracken, most are usually present and exceptions are difficult to find. Pine needles are a particularly selective substrate and of these fungi only *Aureobasidium pullulans* is ever at all common. In the tropics, *Alternaria alternata* is less common and is replaced by *Nigrospora* spp. especially *N. sphaerica*, and *Curvularia* spp., especially *C. lunata*, as is evident from examining senescent leaves of guinea grass (*Panicum maximum*) and banana (*Musa sapientum*). The differences are also reflected in the comparison of the dry air-spora of tropical and temperate climates. The association of these particular fungi has been noted on other substrates, such as cereal stubble and cotton fabrics exposed to the weather. The blackening of the ears of cereals in a damp season is caused mainly by *Alternaria, Cladosporium* and *Epicoccum*. Christensen and Kaufmann (1965), in their studies on the deterioration of grain, designated these and others, such as *Chaetomium*, Fusarium and *Rhizopus* spp., as 'field fungi'. This is an appropriate term as they are almost always and constantly associated with exposed freshly decaying green parts of plants. On leaves they are usually associated with one or more other saprotrophs which are more restricted in the range of leaves which they colonize. These restricted primary saprotrophs may be confined to a particular host genus or a related group of plants.

*Readeriella mirabilis* and *Piggotia stellata* appear to be restricted to *Eucalyptus*. Several species of *Leptosphaeria*, such as *L. microscopica*, are restricted to the Gramineae and *Fusicoccum bacillare* and *Sclerophoma pithiophila* are both very common on pine needles but the latter, at least, is also found on other coniferous leaves. In many of these substrate specificity might be synonymous with, and explained by, host specificity. Many of these, although very active saprotrophs, may have an additional advantage in that they can gain access as parasites. *S. pithiophila*, for instance, has been associated with the defoliation of the current year's needles of *Pinus sylvestris*.

As a group these common primary saprotrophs may be well-established in leaves long before leaf-fall. For example *Cladosporium herbarum* often colonizes and produces conidia on damaged necrotic parts of beech leaves in June, within two months of their unfolding. The duration of their persistence on leaves once they are in the litter is dependent upon many variables, one of which is the texture and another the composition of the leaves. In general, tree leaves, such as those of ash and sycamore, which decompose and disappear rapidly from the litter, support a more substantial growth of these common primary saprotrophs for a shorter time than do leaves of beech and oak which persist much longer in the litter. On beech leaves, for example, *C. herbarum* persists in high frequency through the winter after leaf-fall until the following June and disappears after September.

Similar sequences can be found on other substrates. Primary saprotrophs are the first fungi to appear on flowering stems of cocksfoot, *Dactylis glomerata*. They are present on the basal leaves in early summer and progress up the stems as successive leaves senesce. They are well-established by July and August on the upper leaf sheaths and internodes of stems which flowered in late May and June and they persist there until the following summer. On nettles, *Urtica dioica*, primary saprotrophs colonize the upper leaves in August or September of the year of flowering, at the onset of basipetal senescence. They again persist throughout the winter until the following spring and summer.

## ATTRIBUTES OF THE COMMON PRIMARY SAPROTROPHS

The intriguing aspect of this particular facet of fungal ecology is to ponder why so few of all the fungi are equipped to assume this role of primary saprotrophic colonizers of such exuberantly plentiful substrates.

**Nutrients**

In the well-known schema for fungal successions proposed by Garrett (1963), the primary saprotrophs to invade are 'sugar fungi'. They are non-cellulolytic and rely upon readily available sugars, such as hexoses and pentoses, and other carbon sources simpler than cellulose, such as pectins and starch. These fungi also normally possess a high mycelial growth rate and a capacity for rapid spore germination. The classic example of such fungi is the Zygomycete Mucorales, common on herbivore dung. Primary saprotrophic sugar fungi are usually very ephemeral because of the transient nature of their substrate. The persistence of the common primary saprotrophs for months on leaves would suggest that they are not confined to such ephemeral substrates. The ability to utilize cellulose is often regarded as essential for saprotrophic fungi and the majority, except most Zygomycotina and Mastigomycotina, can do this. Of this particular group of leaf saprotrophs only *Aureobasidium pullulans* is non-cellulolytic. It probably relies on pectic substances for its carbon sources and this ability is often used to explain its role as a primary colonizer. None of the others is markedly cellulolytic when compared with some of the Basidiomycotina which later colonize leaves in the litter layer. For instance, Hering (1967) inoculated oak leaves sterilized by $\gamma$ irradiation with *A. pullulans, Cladosporium herbarum* and *Mycena galopus*, a Hymenomycete agaric from the leaf litter, and measured loss in mass after six months at 9–15°C. The two former brought about a loss of 2 and 4% respectively and the latter 15–20%. Not all the loss in mass was of cellulose but in the latter case the loss corresponded with the utilization of about one sixth of the total cellulose present. On filter paper cellulose, various isolates of *Alternaria alternata* brought about losses of 4–8% in 14 days and *Epicoccum purpurascens* about 4%. For comparison, under the same conditions the vigorously cellulolytic *Chaetomium globosum* brought about a 10% loss. It must be remembered that mass loss methods measure only the amount of substrate, in this case cellulose, respired and lost as carbon dioxide and water and not the amount incorporated into fungal material. The cellulolytic ability of these fungi thus varies and they all probably use simpler carbohydrates, such as sugars and starch, as long as they last and then go on to utilize cellulose even if to a limited extent and slowly. Thus they persist. This may be placing undue emphasis on their carbohydrate nutrition to the neglect of their nitrogen requirements. The nitrogen supply might be extremely critical in determining their distribution. In culture they can all use nitrate, ammonia or amino acids as their sole nitrogen source but nothing is precisely known as to what sources are available to them within the leaf. Indications of the over-riding limitations of their nitrogen supply are seen when leaves are amended with an available source. Foliar applications of 5% urea solution, after harvest but before leaf fall, prevent ascocarp development in the apple scab fungus, *Venturia inaequalis*, on the overwintering leaves and is used as a control measure to limit the ascospore inoculum available to infect the newly emerging leaves in the following spring. Birchill and Cook (1971) in studying the mode of action of the urea demonstrated that both chemical and microbial changes occurred in the leaves after treatment Marked alterations occurred in the composition and density of fungal and bacterial populations present on treated leaves. The urea in particular enormously increased the relative abundance, as assessed by the number of conidia produced, of both *Cladosporium* spp. (Fig. 5.6) and *Alternaria* spp. So marked was the development of the conidia and conidiophores of *Cladosporium* that they could be seen with the naked eye as olive green

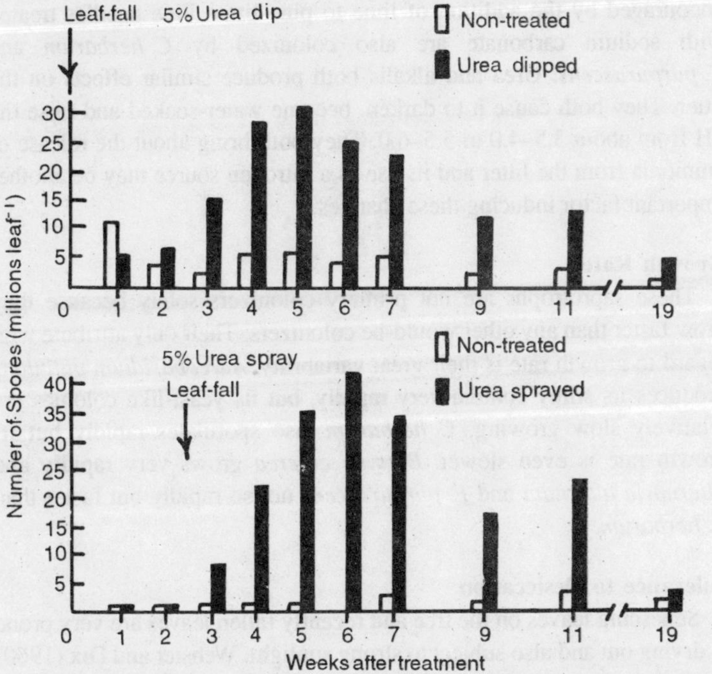


**Fig. 5.6.** Effect of urea dip and urea spray applied 17 October 1967 on numbers of *Cladosporium* spores washed from overwintering leaves.

lawns. Many times more conidia were produced overall on the treated leaves suggesting that the added nitrogen enabled them to utilize more carbon sources and thus outcompete *Venturia* for substrate. Application of urea to fallen pine needles also dramatically changes the fungal succession. *Cladosporium herbarum*, rather than being an occasional inhabitant, is again stimulated to develop to such an extent that its conidiophores may cover the needles as a dense felt and other common primary saprotrophs, such as *Epicoccum purpurascens*, which does not normally occur on pine needles, are stimulated to develop by the urea. The mode of action of the urea is not known in this case but is complex; its property of acting as an alkali may be one important aspect of its effect. For example, several agarics which have not been recorded from pine litter appear when plots are treated with urea or alkalis. An example is *Myxomphalia maura* which is characteristically found on the alkaline ash of bonfire sites on acid soils in coniferous woods. It is not found on woods on alkaline soils so is not a calcicole. *M. maura* is markedly encouraged by the addition of lime to pine litter. Pine needles treated with sodium carbonate are also colonized by *C. herbarum* and *E. purpurascens*. Urea and alkalis both produce similar effects on the litter. They both cause it to darken, become water-soaked and raise the pH from about 3.5–4.0 to 5.5–6.0. They both bring about the release of ammonia from the litter and its use as a nitrogen source may be another important factor inducing these changes.

### Growth Rates

These saprotrophs are not primary colonizers solely because they grow faster than any other would-be colonizers. Their only attribute with regard to growth rate is their great variability. *Aureobasidium pullulans* produces its slimy conidia very rapidly, but its yeast-like colonies are relatively slow growing. *C. herbarum* also sporulates rapidly but its growth rate is even slower. *Botrytis cinerea* grows very rapidly and *Alternaria alternata* and *E. purpurascens* not so rapidly but faster than *C. herbarum*.

### Tolerance to Desiccation

Senescing leaves on the tree and recently fallen leaves are very prone to drying out and also subject to strong sunlight. Webster and Dix (1960) compared the growth rates, latent period for germination, germ tube growth rate at 100% RH and the lowest RH at which spore germination occurred in three primary colonizers with two later secondary colonizers,

*Torula herbarum* and *Tetraploa aristata*. They found that there was little difference between the capacity of the mycelium of the various colonizers to grow at low humidities and the primary colonizers did not make better growth at low humidities. But it can be seen from Table 5.2 that under favourable humidities (100% RH), *A. alternata* and *E. purpurascens* not only grew faster than the secondary colonizers but also had a shorter latent period before germination and their germ tubes grew faster. These features, coupled with the fact that their conidia can germinate at lower relative humidities, would give them an advantage over the others in that their conidia would germinate under less ideal conditions of humidity and they would quickly exploit, by virtue of their more rapid growth rate, any changes to more humid conditions.

**TABLE 5.2**

**Mycelial Growth Rate, Latent Period for Germination, Growth Rate of Germ Tubes at 100% RH and Lowest RH at which Spores Germinated**

| | *Mycelial growth rate (mm day$^{-1}$)* | *Latent period (h)* | *Germ tube growth rate (μm h$^{-1}$)* | *Lowest RH at which germination occurs* |
|---|---|---|---|---|
| **Primary colonizers** | | | | |
| *Cladosporium herbarum* | 2.96 | 6–12 | 4.2 | 89% |
| *Alternaria alternata* | 6.41 | 3–6 | 29.1 | 89% |
| *Epicoccum purpurascens* | 6.45 | 0–3 | 31.6 | 92% |
| **Secondary colonizers** | | | | |
| *Toruia herbarum* | 1.41 | 12–18 | 0 | Water |
| *Tetraploa aristata* | 3.34 | 12–18 | 2.4 | 98% |

Because of the rapidly fluctuating conditions on the leaf surface, germinating conidia may rapidly dry out before penetrating into the leaf. Diem (1971) has investigated the survival at low humidities of germinating conidia of *C. herbarum, A. alternata* and some casual inhabitants of the phylloplane. He found that germinating pigmented conidia, such as those of *Cladosporium* and *Alternaria*, were more resistant than germinating colourless conidia of *Aspergillus* and *Penicillium*. The germ tubes of *Cladosporium* were remarkedly resistant. Some 90% grew on at 100% RH after 8 h in a desiccator over anhydrous calcium chloride and 99% did so after being kept at 40% RH for 8 h. This would indicate that if they germinated in the more humid conditions of the night and had not penetrated into the leaf by the morning they could survive the drier conditions of the day. Germ tubes of the conidia of *Alternaria* were equally resistant but some failed to grow on after periods at relative humidities below 65% but the conidia either germinated again from another cell or from a lateral branch from below the damaged part of the germ tube. In contrast, the germ tubes of the conidia of *Aspergillus* and *Penicillium* were no longer viable after periods at 85% RH. Species with coloured conidia are thus more likely to be successful in the phylloplane and as subsequent primary colonizers of the leaves. But it should be noted that the conidia of *Aureobasidium* and *Botrytis* are not pigmented. The biotrophic Erysiphales which produce their mycelium on leaf surfaces also have colourless hyphae and conidia. Thus pigmentation is a useful but not an essential attribute to possess.

Primary saprotrophs also show an equally remarkable tolerance to desiccation in their hyphal tips as distinct from germ tubes. Hyphal tips are very delicate structures but in these fungi they survive periods of extreme desiccation, some as long as three weeks, above a saturated solution of potassium nitrate ($a_W$ 0.45). Thus they can rapidly exploit the return to favourable conditions of humidity with no apparent loss of previously synthesized biomass. Other fungi do not show this ability. This attribute may be critical in enabling such fungi to tolerate cycles of wetting and drying.

**Survival Structures**

Once established on the leaf surface, most of the common primary saprotrophs produce some form of pigmented survival structure: *Cladosporium herbarum* minute micro-sclerotia; *Botrytis cinerea* and *Epicoccum purpurascens* sclerotia; and *Aureobasidium pullulans* aggregates of chalmydospores (Fig. 5.2b). All have a pigmented mycelium. Such structures and pigmentation protect against desiccation, ultra-violet light and microbial lysis.

It is thus clear that the common primary saprotrophs possess a multiplicity of attributes by which they have become successfully adapted to this relatively inhospitable niche with each fungus possessing its own particular complex of attributes, not all necessarily the same.

As suggested, in temperate climates some of the common primary saprotrophs produce ascocarp initials in the late autumn in the year of leaf-fall as do a number of leaf pathogens, such as *Apiognomonia errabunda* on beech and *Venturia inaequalis* on apple. Ascospores are discharged from these over the period early April to early June. This is the time when the next crop of leaves is unfolding. The initially spore-free leaves become impaction sites for air-borne ascospores and under favourable conditions infection occurs. Such a life history is of particular significance in leaf pathogens with restricted periods of spore formation and release and where the host virtually frees itself of infection by shedding all its leaves prior to its dormant season. The requirement for an overwintering phase, a period of low temperature (5–8°C), before ascocarp initials mature is very common in these fungi. The maturation and release of the ascospores thus coincide with the breaking of bud dormancy of the host. Thus teleomorphic states of *Aureobasidium pullulans (Guignardia fagi)* and *Cladosporium herbarum (Mycosphaerella tassiana)*, which are common on fallen leaves, may be regarded as additional survival structures adding an ascospore inoculum to the conidial inoculum available in the spring.

**Subsequent Colonizers and Leaf Decay**

These initial colonizers gradually disappear, being replaced by other leaf-inhabiting saprotrophs which begin to reproduce in the late summer of the year after leaf-fall, reach a maximum in the autumn and persist over the winter, until the spring. These include a very wide variety of conidial fungi, such as *Polyscytalum fecundissimum* and *Chalara cylindrospora*, Ascomycotina such as *Microthyrium fagi*, and *Helotium caudatum*, and Basidiomycotina with minute basidiocarps, such as *Lachenlla villosa* and *Pistillaria pusilla* on beech leaves. With fragmentation in the final stages of decomposition, the fungal flora becomes dominated by typical soil-inhabiting fungi, mainly Zygomycete Mucorales, especially species of *Mucor* and *Mortierella*, and conidial fungi, such as species of *penicillium* and *Trichoderma*, together with litter-decomposing Hymenomycete Agaricales, such as species of *Collybia* and *Mycena*.

The soil-inhabiting fungi grow up from the soil via the continuum of organic debris. The role which they play in the decomposition process has not been fully elucidated. At this stage the Mucorales are certainly not using any simple carbohydrates initially present in the leaves as these would have been utilized already. They could be living in association with the cellulolytic Agaricales as commensals by taking a share of the hydrolytic products of cellulose and thus acting as secondary saprotrophic sugar fungi rather than primary ones. Alternatively, they could be primary colonizers of the wealth of faecal pellets produced by the micro-fauna, especially mites, as they are on pellets of *Glomeris*. The H layer of the soil is particularly rich in chitin in the form of hyphal wall fragments and exoskeletons of insects or other chitinized remains of the micro-fauna. Species from several common general of soil-inhabiting fungi, including *Mortierella*, *Penicillium* and *Trichoderma*, have the ability to break down this very resistant substrate and their activity may well represent one of the final stages of the mineralization of primary and secondary organic materials in the soil.

## DECOMPOSITION OF PINE NEEDLES

The time period between leaf-fall and the final decomposition of a leaf varies enormously. In cool North temperate pine forests it may be 10 years or more, in ash and sycamore under 1 year and in tropical forests mere weeks. Pine needles are extremely durable and decay very slowly. Their decomposition most often results in the formation of a mor type of soil. The needles are shed mainly in August and September and there is an accumulation of considerable bulk of leaf litter, each successive leaf-fall burying the previous one so that a stratified litter layer is produced. The animals in the litter are sufficiently small not to disturb this stratification and such a litter layer well illustrates the diversity of organisms, fungi and animals, involved in the decomposition process. A very considerable amount of potential energy is available for micro-organisms in this litter. In *Pinus sylvestris*, production of needles accounts for about one third of the total productivity and accounts for about 60–80% of the total litter. Although the decay process is a continuous, if fluctuating, one, it is convenient to recognize a number of stages. The A horizon may thus be divided into L, $F_1$, $F_2$ and H layers. The L layer consists of freshly fallen, undecomposed needles, light brown to buff in colour and others somewhat darker in colour, which have fallen earlier. Needles remain in this layer for about six months. They all have a high tensile strength, a relatively low but fluctuating moisture content and form a loose, uncompacted layer on the litter surface. In this layer the needles are very susceptible to drying out and conditions are unfavourable for continuous fungal growth. In the upper parts of the $F_1$ layer, the needles are grey, becoming dark brown with depth but recognizable as needles. Their tissues become softened and they have a low tensile strength and a high moisture content. They remain in this layer for about two years. Below, in the $F_2$ layer, the character of the needles again changes. They are greyish, fragmented and compressed, but again still recognizable as needles. The mesophyll collapses and most bear dark amorphous faecal masses of the micro-fauna. Eventually the remains of the needles enter the H layer which consists of an amorphous mass of faeces and the remains of both the micro-fauna and fungi, the needles having undergone complete physical reduction. Below this layer is an intimate mixture of humus and mineral soil.

Two factors that may greatly influence the sequence of decomposer fungi on the needles are the time at which they fall and their previous history. Pine needles are far from being a homogeneous entity and at needle fall vary in age, physical structure, nutrient content and the presence or absence of fungal colonizers in or on the needles. The needles have a very thick, waxy cuticle and support a much sparser population of phylloplane inhabitants. *Sporobolomyces roseus*, although present on most attached needles, occurs in very low frequencies only. This contrasts markedly with its abundance on leaves of deciduous trees and herbaceous plants. It decreases rapidly on needle fall, whereas some other yeasts, such as *Bullera* spp., increase in frequency and persist. A number of other mycelial fungi such as the conidial *Sclerophoma pithiophila* may grow and sporulate on the leaf surface. Vigorous ones, such as *Lophodermella sulcigena*, an apothecial Ascomycete, and *Coleosporium senecionis*, a rust, cause premature needle cast, either directly or by predisposing first year needles to infection by secondary pathogens. For example, *L. sulcigena* infects young first year needles and predisposes them to infection by *Hendersonia acicola* or *Lophodermium pinastri* and finally *Naemocyclus niveus*, which cause the needles to fall in their first summer. Such weak pathogens may colonize the needles directly but spread very little until senescence. They may also gain access via tissues damaged by insect pests. Living needles may also be colonized by *Fusicoccum bacillare* or *Sclerophoma pithiophila*. Needles infected by either of these two conidial fungi soon die and turn brown but remain attached to the tree. Such needles again fall in the summer. *S. pithiophila* is also a frequent colonizer of needles containing high nutrient levels, such as first year needles, shed while still green and of needles of felled pines. Clearly pine needles can fall at varying times of the year and may already be colonized by a variety of fungi which have already initiated the process of decomposition. *Lophodermella sulcigena* actively decomposes the mesophyll tissue and *Hendersonia acicola* may remove much of the cellulose, reducing the needle to a skeleton of epidermal waxes and lignified tissues. *Lophodermium pinastri* produces pigmented diaphragms across the needles delimiting the extent of its colonization (Fig. 5.7a). Such parts later escape extensive internal attack by saprotrophic needle-inhabiting fungi. This is often attributed to their inability to penetrate the melanized diaphragms but in culture at least *Lophodermium* produces powerful antifungal antibiotics and they may also play a part in restricting saprotrophic colonization. Such parts of the needles decay more slowly than uninfected parts and as a consequence accumulate in lower layers of the litter. The saprotrophic colonization of naturally fallen needles and needles shed after parasitic attack may thus be distinct.

Most needles, the bulk being second and third year ones, falling in August and September, are colonized by *L. pinastri* and somewhat fewer by *S. pithiophila*. Soon after needle-fall, a dark brown to black hyphal network develops on the surface of the needles (Fig. 5.7). A number of fungi may be involved, including the conidial *Sympodiella acicola* and *Helicoma monospora* (Fig. 5.7 b and c) and, in drier situations, the ascocarpic *Kriegeriella mirabillis* (Fig. 5.7 b and d). There is no apparent penetration of the needles by the surface hyphae although erosion of the needle surface does occur. The hyphal network shows marked linearity with the hyphae growing longitudinally along the cell boundaries. Internally the needles become colonized by *Desmazierella acicola*, which produces its conidial state from compacted, pigmented, hyphal cushions formed over the stomata. In spite of intensive grazing of the fungi by the micro-fauna, including mites, springtails and enchytraeid worms, all become more frequent as the needles become incorporated into the more moist regime of the $F_1$ layer. They persist for two years in the $F_1$ layer, that is for up to 2 1/2 years after needle-fall. *D. acicola* produces crops of conidiophores in both the first and second summers after needle-fall. *L. pinastri* produces its ascocarps in the L layer over the period January to May, providing the inoculum to infect further needles on the trees. After about 10 months in the L and $F_1$ layers, it too disappears. In the $F_2$ layer, which the needles enter in the third year after needle-fall, the micro-fauna assume more importance. The external feeders continue to graze upon the fungal hyphae and reproductive structures whilst the internal feeders rapidly comminute needles attacked by *L. pinastri* and *D. acicola*. Any needle fragments which escape extensive internal colonization become colonized by more general litter inhabitants such as species of *Penicillium* and *Trichoderma* and by pine litter-inhabiting agarics. Needles remain in this layer for about 7 years, by which time the

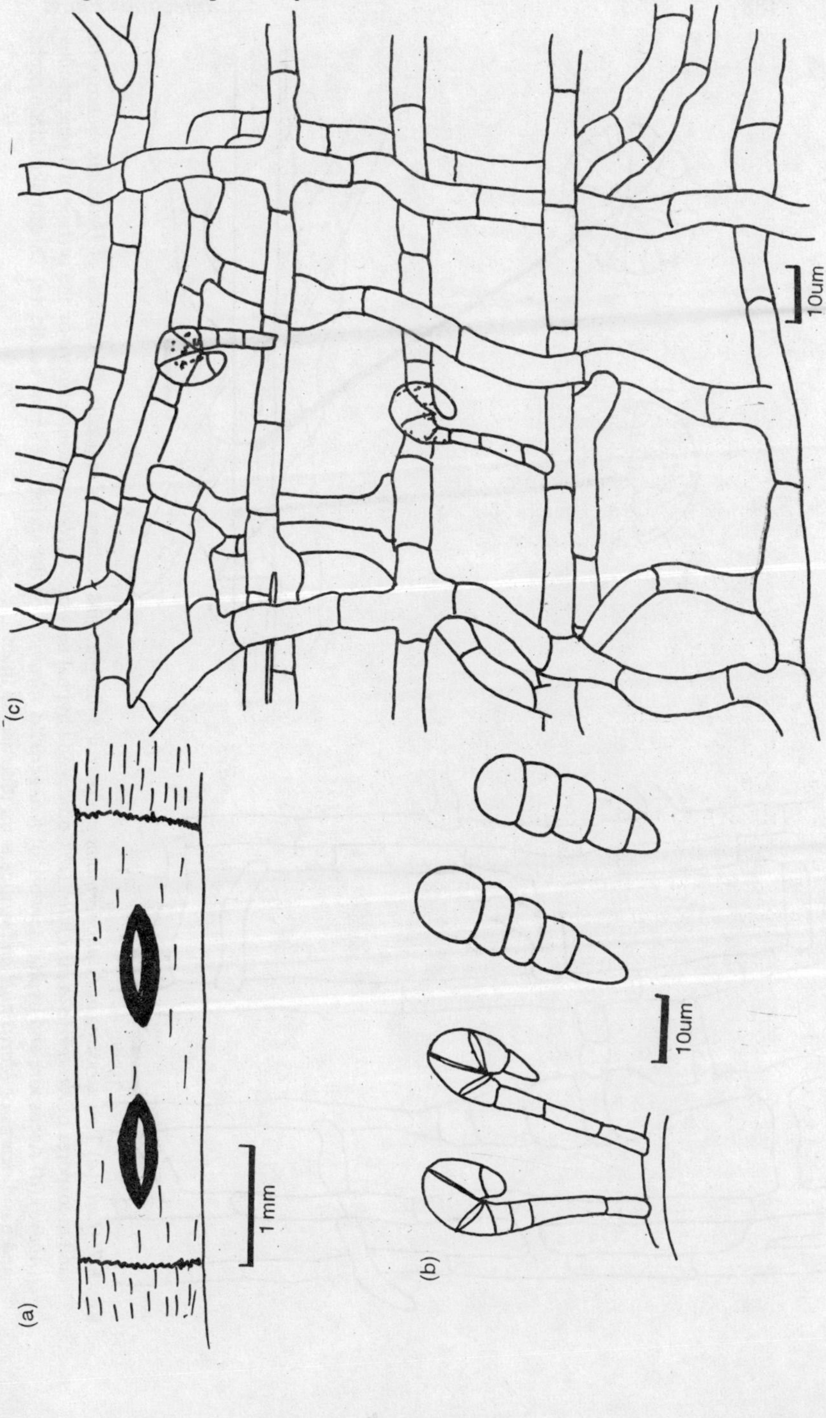
(c)
10um
(a)
1 mm
(b)
10um

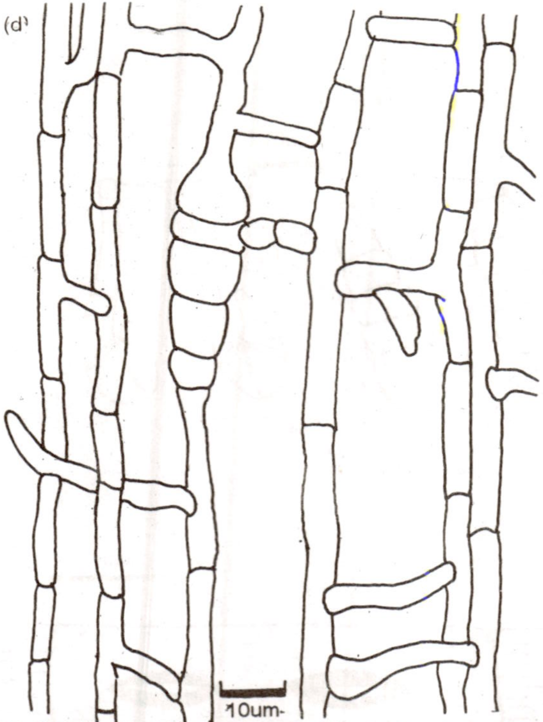


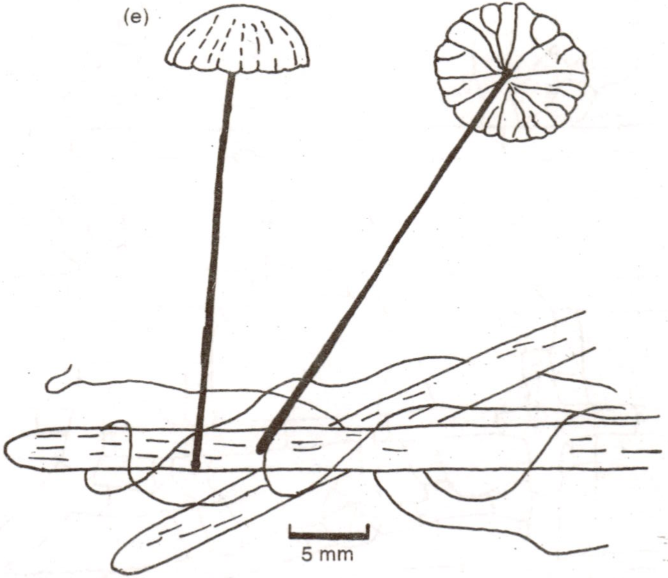


**Fig. 5.7.** (*Left*) (*a*) Two asocarps of *Lophodermium pinastri* and disaphragms across a pine needle. (*b*) Conidia of *Helicoma-monospora* and ascospores of *Kriegeriella mirabilis*. (*c*) Conidia and hyphal network of *Helicoma monospora* on the surface of a pine needle. (*Above*) (*d*) Ascospore and hyphal network of *Kriegeriella mirabills*, on the surface of a pine needle. (*e*) Thread like thizomorphs and basidiocarps of *Marasmius and rosaceus* on pine needle litter.

fungi and fauna reduce them to an amorphous mass, typical of the humus layer. The role of agarics in the decomposition of pine needle litter has not been extensively investigated. It is usually assumed that they colonize the litter when it is in a relatively late stage of decay. This is not always so. The tiny agaric, *Marasmius androsaceus*, is very common in pine needle litter. It is often called the 'Horse hair fungus' because its stalk is shiny black, like horse hair and is of about the same diameter. *M. androsaceus* colonizes the needles very shortly after needle-fall. Its delicate black, cotton-like rhizomorphs grow up from previously colonized needles below, binding them together in a loose tangle. Dense masses of basidiocarps may appear on the needles in the litter, any time from May to November (Fig. 5.7). It is both strongly cellulolytic and ligninolytic and causes very extensive internal decomposition. The role of such Basidiomycotina should not be underestimated. Their mycelium is often prolific in both the L and F layers. Long lists of agarics have been recorded from pine woods. Richardson (1970) has estimated the total productivity of these in a woodland of *Pinus sylvestris* in Scotland, to be between 0.25–0.5 million basidiocarps $10^4 m^{-2} y^{-1}$. The majority are produced from August to September. However, because of our inability to distinguish species of Basidiomycotina from their mycelium and because we know insufficient about the biology of some of these, the problem is to assess the relative contributions of the litter decomposers and the mycorrhizal fungi. Since many agarics in the litter decompose both cellulose and lignin, it is probably delignification that reduces the needles to a greyish colour in the $F_2$ layer.

## THE ROLE OF THE LITTER MICRO-FAUNA

As in other litter systems, the micro-fauna are important agents in the decomposition process. Mites and springtails cause considerable comminution of the needles and in so doing convert them to faecal pellets. It has been estimated that a pine needle with a surface area of 180 $mm^2$ would have a surface area of 1.80 $m^2$ after comminution to faecal pellets by micro-arthropods. Such comminution would present a much larger surface area to microbial enzymes and thus be expected to increase the decomposition rate. This may not always be so. Orobatid mite pellets persist longer than the source from which they are derived. This may be due to the nature of the substances cementing the particles of the pellets together and digestion by the animal of the more easily decomposable components of the litter. It is clear, however, that the water holding capacity of the pellets is higher and the rate of evaporation from the pellets decreased. This creates a higher and more stable moisture regime which again should favour microbial activity.

The enchytraeid worms, which are very abundant is podsols in northern coniferous forests, in addition to their grazing activities, play a vital role in the absence of earthworms in this more type litter, in mixing the amorphous remains with the mineral soil. The major group of animals involved in the decomposition of pine needle litter are mites, many of which are strictly mycophagous, with some showing marked preferences for particular fungi. They are most abundant in the moisture F layers where fungi are also more active. Protozoa and nematodes also occur in pine litter and feed on the contents of living fungal hyphae. There is no clear evidence to support or refute the hypothesis that mycophagy by the micro-fauna stimulate the growth of fungal mycelia.

Perhaps the most important effect of the micro-fauna is that they act as a reservoir of plant nutrients as they do in decaying wood, gradually making available the minerals which have become immobilized in the fungal hyphae. The large number of different species involved and their varying life spans mean that the nutrients contained in their tissues are only gradually mineralized.

# 6

# FUNGI AS DECOMPOSERS OF WOOD

Perennial woody plants are the predominant vegetation on earth. Forests form the climax vegetation of all parts of the world except where temperature and moisture extremes limit plant growth. Forests also contain the greatest biomass varying from 500 Mg $10^4$ $m^{-2}$ in tropical rain forests to 100–300 Mg $10^4$ $m^{-2}$ in northern temperate coniferous forests. Perennial woody parts above ground make up about three-quarters of this biomass. Woody tissues thus provide the bulk source of organic carbon for decomposer heterotrophs. Fungi are the major group of organisms responsible for wood decay and a number of groups of fungi are solely wood-inhabitants. They exist entirely on the components of wood. In this chapter, a detailed consideration of these fungi will further illustrate the versatility of fungi as saprotrophs.

## THE STRUCTURE AND COMPONENTS OF WOOD

Technically, wood is the xylem cylinder inside the bark of trees. In many trees it consists of an outer, light coloured sapwood and an inner, darker heart-wood. The bulk of the wood consists of dead and empty lignified vessels and/or tracheids and fibres but it also contains xylem parenchyma. Much of the parenchyma in the sapwood remains alive and unlignified and acts as a food store, mainly for starch but soluble sugars, proteins, peptides, and amino acids, lipids, nucleic acids and vitamins, such as thiamine, are also present. Once the tree is dead, these afford substrates for a wide variety of fungi but are all relatively minor and ephemeral components.

Wood consists of three major components – 40–60% cellulose, 10–30% hemicelluloses and 15–30% lignin. Although the biological decomposition of lignin is of critical importance in the continuous cycling of carbon, its degradation is incompletely understood. This can be attributed to many facts; not least of these are our lack of understanding of its precise chemical structure, the diversity of its structure in different woods, our inability to produce a pure form of lignin for cultural studies and a suitable assay for lignin degradation, the availability of a very potent lignin degrader, and the general cellular and chemical complexity of wood.

Lignin, in addition to making up about one-quarter of the dry mass of wood, is undoubtedly the structurally most complex of all the polymers and the most resistant of all to microbial decomposition. It is a three dimensionally branched aromatic polymer, formed by the oxidative polymerization of three different building blocks, not just one as in cellulose. The building blocks are the phenyl propanes: coumaryl, coniferyl and sinapyl alcohol (Fig. 6.1a). The lignin of different plants may contain different proportions of the three building blocks. Conifer lignin consists of mostly coniferyl alcohol, with small amounts of coumaryl alcohol and minor amounts of sinapyl alcohol. In angiosperm lignin there are approximately equal amounts of coniferyl and sinapyl alcohol and minor amounts of coumaryl alcohol. These phenyl propane units are built up into a branched polymer by covalent bonding involving three major linkage types. By far the commonest, making up 40–60% of the total, and most important linkage type is the arylglycerol-β-aryl ether type. Phenylcoumaran structures form 10–20% of the linkage types and biphenyl structures another 10–25% (Fig. 6.1b). Thus in lignin there are three functional monomers, varying in proportions in the various lignins, and three major linkage types, but also other minor ones. There is no regular repeating unit as there is in starch or cellulose, nor are there bonds which are easily hydrolysed. Because of this structural complexity, decomposition must necessarily differ from that of most natural polymers where there is usually straight cleavage, often by hydrolysis, to produce the monomers. There is also the possibility of microbial enzymes bringing about a variety of limited changes to the intact molecule and only partially degrading it to substances which pass with little further change into humic materials. It appears that only the so-called white-rot fungi can completely decompose lignin to carbon dioxide and water. Lignin

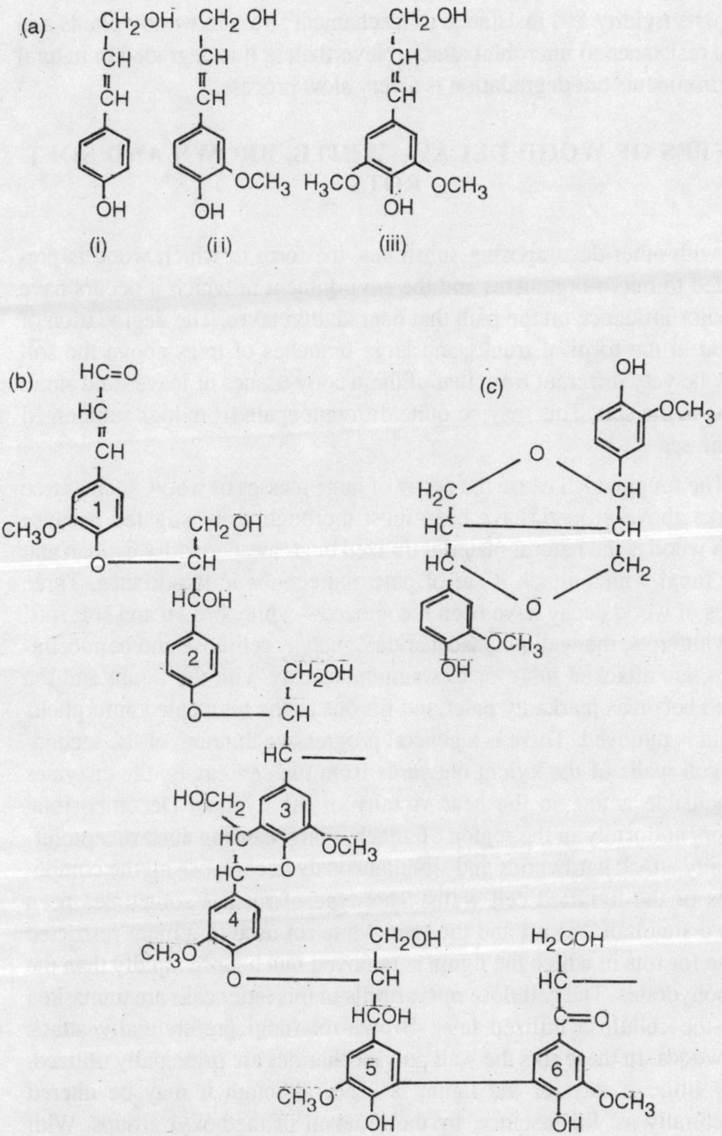

**Fig. 6.1.** (*a*) Building units of lignin: (*i*) coumaryl; (*ii*) coniferyl; and (*iii*) sinapyl alcohol. (*b*) Major linkage types of lignin. (*c*) Pinoresinol, a dilignol.

imparts rigidity and resistance to mechanical stress in woody plants and also resistance to microbial attack. Nevertheless it is degraded in natural environments but degradation is a very slow process.

## TYPES OF WOOD DECAY—WHITE, BROWN AND SOFT ROTS

As with other decomposing substrates, the form in which wood is presented to micro-organisms and the environment in which it occurs have a major influence on the path that degradation takes. The degradation of wood in the form of trunks and large branches of trees above the soil may be very different from that of the woody tissues of leaves and small roots in the soil. This may be quite different again from logs submerged in the sea.

The fungi which cause the decay of large masses of wood, such as tree trunks above ground, have been most thoroughly investigated because such wood is the natural material utilized in greatest quantity by man and any fungi which attack it are of potential economic importance. Three types of wood decay have been recognized—white, brown and soft rots. In white rots, the wall polysaccharides, such as cellulose and hemicelluloses, are attacked more or less simultaneously with the lignin and the wood becomes markedly paler and fibrous as the pigmented amorphous lignin is removed. There is a general progressive thinning of the secondary cell walls of the xylem outwards from the cell cavity, the enzymes responsible acting in the near vicinity of the hyphae. Decomposition occurs uniformly in the region of attack. Fungi causing such rots preferentially attack hardwoods and simultaneously decompose all the components of the lignified cell walls. This type of rot has sometimes been called simultaneous rot and the term white rot used in a more restricted sense for rots in which the lignin is removed much more rapidly than the carbohydrates. The cellulose microfibrils in this latter case are unmasked and the cellulose utilized later. Brown-rot fungi preferentially attack softwoods. In these rots the wall polysaccharides are principally utilized. Very little, if any, of the lignin is used, although it may be altered structurally as, for instance, by the removal of methoxyl groups. With decay the wood becomes darker brown. There is no thinning of the walls. The enzymes responsible diffuse away from the hypha and act on the entire cell wall, often at some distance from the hyphae. The structural polymers are removed, leaving a framework of lignin to maintain the general cell shape so that there is little apparent damage until the cell walls collapse. Decomposition occurs in irregular patches in the attacked wood. This leads to the cubically cracked appearance of brown-rotted wood. It also crumbles readily to a powder when rubbed between the fingers. In both white and brown rots, the hyphae grow and branch in the cell cavities and penetrate the walls mechanically via pits or the surfaces in general, by coupling penetration with enzymic erosion, producing bore holes somewhat wider than the hyphae. In both, they penetrate deeply into the wood. Soft rots, on the other hand, are more conspicuous near the surface, advancing inwards after destroying the outer layers of the wood. They occur only in wood of unusually high moisture content, such as water-logged river and marine timbers. The soft-rot fungi again principally utilize the cellulose and the hemicelluloses of the walls, but their hyphae penetrate and grow within the secondary cell walls; here they enzymatically create chains of typically rhomboidal or elongated cylindrical cavities, with conically tapering ends (Fig. 6.2). Decomposition is restricted to the immediate neighbourhood of the hyphae.

Whereas soft rots are caused by Ascomycotina, such as species of *Chaetomium* and *Ceratocystis*, and anamorphic states such as *Alternaria* and *Phialophora*, white and brown rots are caused mainly by Basidiomycotina. Two good examples of these are *Coriolus versicolor* and *Piptoporus betulinus* respectively. The former is one of the commonest polypores and is found on a great variety of hardwoods whereas the latter is a facultative wound parasite restricted to birch (*Betula* spp.). *Coriolus* can degrade over 90% of the lignin in wood. Several hundred species in the Hymenomycete Agaricales, but more so in the Aphyllophorales, an order almost entirely confined to wood, cause white rots. But apart from these, only a very few Ascomycotina, including *Xylaria polymorpha* and *Ustulina deusta*, can cause a white rot. Somewhat fewer Basidiomycotina cause brown rots. Thus even in the fungi, the ability to degrade lignin completely is limited to the relative few.

## LIGNIN DEGRADATION

It is still not at all clear how these fungi act on lignin, in spite of the fact that many studies have been made on the effects of white-rot fungi on lignin model compounds consisting of two phenyl propane units, such as the dilignol pinoresinol (Fig. 6.1c), or extracted lignin in liquid culture and the changes that occur in wood as it rots. Extracted lignins have been of little use in laboratory studies simply because of the physical and chemical changes brought about in its structure on extraction. Two

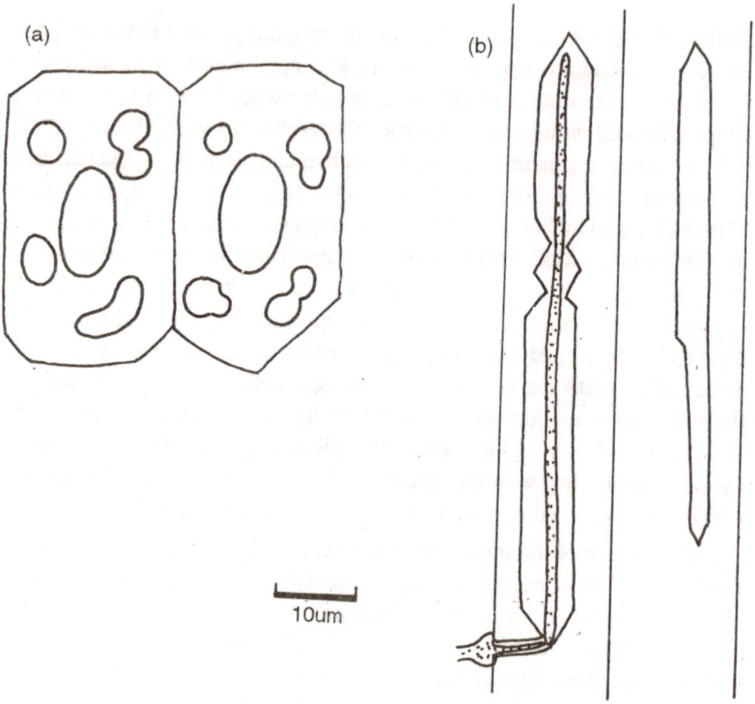


**Fig. 6.2.** Diagram of (*a*) transverse and (*b*) longitudinal sections of tracneids of *Pinus* with soft rot cavities in the secondary walls.

extracted lignins, Kraft lignin and lignosulphate, have been widely studied because they are produced in such vast amounts as waste products of the paper industry. The compound which has been most widely used is a synthetic lignin designated DHP (dehydrogenative polymerizate). Like lignin, it is produced by condensation and it is chemically very similar but has a much lower molecular weight. $^{14}C$ labelled DHP is usually used. The amount of $^{14}CO_2$ evolved is the most sensitive measure of ligninolytic activity.

**Role of Extracellula Phenolases**

When cultured on agar containing phenolics, such as gallic or tannic acid, most—over 90%—of the white-rot fungi produce extracellular phenolases, such as laccase, peroxidase and tyrosinase, which oxidize these acids; a brown coloured diffusion ring appears around the colony margin. These catalyse the removal of electrons from phenols. They have long been considered as being involved in lignin degradation because lignin is a phenolic and so a substrate for these and lignin degradation is certainly an oxidation. Lignin is resistant to decomposition in anaerobic conditions. Also white-rot fungi produce these enzymes but the closely related brown-rot fungi, which do not decompose lignin, do not. Thus there is the apparent correlation between the ability to degrade lignin and the production of extracellular phenolases. The ability to degrade implies the formation of smaller compounds, yet it is usually considered that these oxidases act by coupling and polymerization to form compounds of higher molecular weight. There is, however, some scant evidence from experiments using white-rot fungi and model compounds that these enzymes can bring about limited depolymerization. It has also been shown that the continued action of these enzymes on wood itself leads to some degradation of the lignin. It is doubtful, however, whether they play any significant part in lignin degradation. In any case they can only be part of the enzyme complex which attacks lignin. It has been suggested that they have an indirect role in polymerizing and so detoxifying any toxic phenolics released during degradation—that is to suggest that they act after the monomers have been cleaved off. Simple phenolics are often toxic to fungal growth and they may have an important function in coupling these. This would be comparable to depside and depsidone formation from monocarboxylic acids in lichens.

**Cleavage of Major Linkage Groups**

One obvious step towards decomposition would be to cleave any of the major linkage groups between the phenyl propane units tc release the $C_6$–$C_3$ monomers. *Coriolus versicolor* and a number of other white-rot fungi can cleave lignin models bonded by the arylglycerol-β-aryl ether bond. Although oxidative cleavage of the β-ether linkage occurs, there is again no convincing evidence that any great part of the lignin molecule is cleaved by white-rot fungi to produce the single monomers. There is also evidence that the monomers may be attacked while still bonded in the polymer, not by breaking the bonds between them but by directly attacking the aromatic rings, by either ring cleavage and/or demethylation of the methoxyl groups to hydroxyl ones. Demethylation may also be coupled with side chain oxidation. These are oxidized by the loss of two carbon atoms and the formation of new carboxyl groups. Both these, the formation of –OH and –COOH groups, would lead to increased solubility. Support for demethylation and side chain oxidation comes from two sources. Ligning degraded by white-rot fungi contains less carbon, slightly less hydrogen, fewer methoxyl groups but more oxygen and carboxyl and hydroxyl groups. Culture filtrates from white-rot fungi grown on extracted lignin contain small amounts of vanillin, vanillic acid and syringaldehyde. There is also slightly less vanillin in rooted wood compared with sound wood. This is taken to indicate that some phenyl propane units, either in the lignin or after cleavage, have had their side chains oxidized with the loss of two carbon atoms. Alternatively, it has been argued that vanillin and vanillic acid are attached as side groups along the main polymer and are released on hydrolysis.

### A Hypothetical Scheme for Lignin Degradation

Although white-rot fungi unquestionably can use lignin as a sole carbon source and completely decompose it, we are by no means certain as to how lignin is degraded. A number of very hypothetical schemes have been put forward. These usually assume initial cleavage of the arylglycerol-β-aryl ether bonds between the monomers (Fig. 6.3). This is followed by oxidative cleavage of the side chain with the loss of two carbon atoms and the formation of a carboxyl group to give vanillic acid. Vanillic acid is demethylated to protocatechuic acid. Ring cleavage then occurs to keto-adipic acid and this is used in the tricarboxylic acid cycle. It is most likely that these reactions occur simultaneously over the surface of the polymer with the oxidative cleavage of the side chains being centrally significant for fragmentation of the polymer.

### Role of Agents Other than Enzymes

Evidence is accumulating that agents other than enzymes, such as the hydroxyl radical (OH), may be involved in lignin degradation. In cultures of *Phanerochaete chrysosporium*, one of the most widely used white-rot fungi, hydroxyl dependent formation of ethylene coincides with ligninolytic activity. The radical is probably formed from hydrogen peroxide in the so-called Haber-Weiss reaction which is catalysed by iron and requires the superoxide radical ($O_2^-$)

$$O_2^- + Fe^{+--} \rightarrow Fe^{++} + O_2$$

$$Fe^{++} + H_2O_2 \rightarrow Fe^{+++} + OH^- + OH$$

The dependence of ligninolytic activity on the radical is verified by the fact that specific OH quenchers, such as mannitol, inhibit, it. Wood rotting fungi produce sufficient hydrogen peroxide, by the action of a variety of oxidases, from the components of wood. Such oxidases are

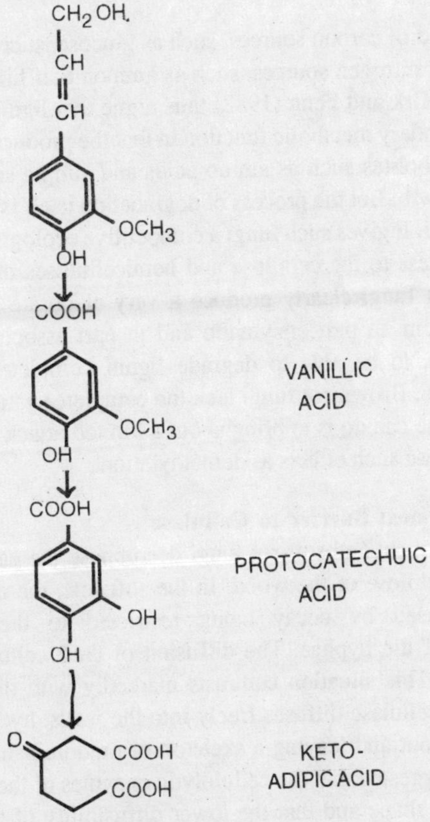


**Fig. 6.3.** Schema for lignin degradation.

synthesized most rapidly when readily available carbon and nitrogen containing nutrients are low, when ligninolytic activity is at its peak. Sufficient amounts of $Fe^{++}$ are also present in wood.

Cultural conditions are critical for ligninolytic activity. For example, to convert some 40% of DHP to carbon dioxide and water with *P. chrysosporium*, the culture must be maintained in the stationary phase at pH 4–5, with very low levels of metabolizable carbon compounds, low available nitrogen and high oxygen concentrations. The oxygen supply is a very critical variable as can be shown from the fact that incubation in pure oxygen increases lignin degradation 10-fold over incubation in air. The rate of degradation also increases if very thin mycelial mats are used rather than thick ones, diffusive supply being important. Cultures have also to be starved of carbon sources, such as glucose, sucrose and cellobiose, as well as nitrogen sources, such as ammonia, if high rates are to be maintained. Kirk and Fenn (1982) thus argue that lignin degradation is a strictly secondary metabolic function in that the products, as opposed to primary metabolites such as amino acids and simple sugars, are not essential for growth. But the process of degradation itself is of a selective value to the fungi. It gives such fungi a competitive ecological advantage in providing access to the cellulose and hemicelluloses masked by the lignin. White-rot fungi clearly produce a very elaborate and complex ligninolytic system, in part enzymatic and in part associated with the hydroxyl radical, to be able to degrade lignin completely to carbon dioxide and water. Brown-rot fungi lack the complete system. The most that many of these can do is to bring about a limited attack on the lignin molecule and cause such effects as demethylation.

**Lignin as a Physical Barrier to Cellulase**

Both the brown-rot and soft-rot fungi decompose the carbohydrates, especially the cellulose of the wood. In the soft rots, the characteristic cavities are caused by decay being restricted to the immediate neighbourhood of the hyphae. The diffusion of their cellulase is definitely restricted. This situation contrasts markedly with the brown-rot fungi where the cellulase diffuses freely into the walls, hydrolysing the cellulose throughout and leaving a skeleton of predominantly lignin. It has often been suggested that the cellulolytic enzymes of the two groups differ in size and shape and that the lower diffusibility of the cellulase produced by the soft-rot fungi indicates larger molecular dimensions but this is not so. The cellulases produced by the two groups have similar dimensions and properties. Many actively cellulolytic fungi may be restricted in their ability to utilize cellulose in wood by virtue of the intimate nature of the association between the cellulose and the lignin. A particularly good example is *Chaetomium globosum* which rapidly degrades cotton and filter paper cellulose completely, but only attacks wood of high moisture content and merely produces soft rot cavities in the cell walls. Lignin appears to act as a physical barrier that prevents the cellulase from reaching sufficient glycosidic bonds in the cellulose to permit any large scale hydrolysis. Thus the accessibility of the cellulose to the degrading enzymes is a most important factor. The evidence for this comes from a number of sources. Increased accessibility can be achieved by breaking down the wood to much finer particles before adding cellulose. This exposes a larger surface area of the cellulose free of its association with lignin. For example, in experiments using sawdust and ball-milled sawdust, increased hydrolysis occurred in the latter when cellulose was added. It thus appears that brown-rot fungi possess some system—a pre-cellulolytic phase—which enables the cellulose to get at cellulose in wood. In the cell walls of wood, the cellulose microfibrils are encrusted with and surrounded by lignin and hemicelluloses. One suggestion that has been made is that brown-rot fungi produce enzymes which the soft-rot fungi lack. Some of these degrade the hemicelluloses and others disrupt the links between the cellulose and the lignin. There is also some evidence that one does not necessarily have to postulate enzymic dissociation of the lignin from the cellulose. Brown-rot fungi growing in wood, develop and maintain their own pH of between 2–4, whereas soft-rot fungi develop best in near neutral conditions. It may well be just that acidic conditions are necessary to disrupt the association between the lignin and the cellulose. If wood is treated initially with acid and, after removing the acid, soft-rot fungi allowed to attack it, the fungi bring about a greater loss than in untreated wood.

## NATURAL RESISTANCE OF WOOD TO FUNGAL DECAY

### Lignification

Lignification of the cell walls is obviously a very important factor that contributes to the natural resistance of wood to fungal decay. This is more important in soft rots. Softwoods are more resistant to these than hardwoods. This is usually attributed to the higher degree of lignification and the higher density of cell walls in conifers. It seems unlikely that mere abundance of lignin can account solely for the difference in resistance. The different proportions of the various phenyl propane units in the lignin, the degree of cross linkage with the cellulose and the nature of the hemicelluloses must also be important. Nevertheless lignification must act as some sort of physical barrier. Many very actively cellulolytic fungi and bacteria cannot attack wood because the lignin prevents their cellulose from reaching sufficient glycosidic bonds to permit hydrolysis on such a scale that they can grow on the proceeds.

### Refractivity of Cellulose

Many other factors contribute to decay resistance. The cellulose in wood tends to have a higher degree of refractivity or crystallinity than in the cell walls of herbaceous plants. The microfibrils are more highly ordered and there are correspondingly less amorphous or more randomly organized areas. The higher the refractivity, the smaller the surface immediately accessible to the components of cellulase.

### Nitrogen Content

In addition to being distinguished by its high lignin content, wood can also be distinguished from other plant materials by its very low nitrogen content. This also increases its resistance to decay. Woody tissues contain 0.03–1.0% nitrogen as compared to 1.0–5.0% in herbaceous tissues. The carbon : nitrogen ratio in most woody tissues is thus high, in the order of 350–500 : 1 and may exceed 1000 : 1. For most fungi a substrate with such a high carbon : nitrogen ratio would be nitrogen deficient and growth limiting. Wood-decay fungi are unusual in that they can grow in such substrates. They metabolize large amounts of carbohydrates (and lignin in white rots) in the presence of very small amounts of nitrogen. The mycelium of most fungi, grown on nutrient media, contains about 5.0% nitrogen and has a carbon : nitrogen ratio of about 10 : 1. The nitrogen content of the medium may fall, under starvation conditions, to around 1.0% before growth stops. The white-rot fungus *Coriolus versicolor* is unusual in that on high carbon : nitrogen containing media, the total nitrogen in the mycelium may fall as low as 0.2% before growth rapidly declines. The ability to grow under such conditions suggests a greater efficiency in its nitrogen metabolism. This may be achieved in a number of ways. Mycelial nitrogen may be re-used either by internal translocation from old to young hyphae or by autolysis and re-use. Extracellular lytic enzymes may be secreted which break down the old hyphal walls making the constituents, especially the nitrogen in the chitin, available for re-assimilation. Preferential allocation of available nitrogen to nucleic acids and enzymes may occur. For example, when growth of *C. versicolor* on media containing low and high carbon : nitrogen ratios was compared, the total nitrogen, expressed as percentage dry mass of the mycelium, fell from 4.4 to 0.2 but the percentage nitrogen in nucleic acids rose from 4 to 25 and the amount of cellulase produced per unit of mycelium was comparable in each. In fungi in general, cellulolysis diminishes with increase in the carbon : nitrogen ratio. White-rot fungi are unique in being able to produce cellulase at a carbon : nitrogen ratio of 2000 : 1 whereas in most other fungi this ability is negligible at a ratio of about 2000 : 1.

**Moisture Content**

Wood-decay fungi have higher moisture requirements for growth than fungi which attack most other plant materials. Their growth rate is very sensitive to changes in the water activity ($a_w$) of the medium. Whereas cotton is susceptible to fungal attack when it has a moisture content of more than 10% on a dry mass basis and cereal grains more than 13%, wood decay can be initiated only at moisture levels of about 26–32%. In standing trees and freshly felled timber, most of the cell cavities in the wood are water-filled. Such wood may have a moisture content of well over 100%. Such completely saturated wood is quite immune to fungal attack, presumably because the oxygen tension is too low to support active hyphal growth and the carbon dioxide content raised considerably above atmospheric levels. Many wood-decay fungi are very tolerant of high carbon dioxide concentrations. Whereas litter-inhabiting Basidiomycotina may be inhibited from growth by a partial pressure of 10 kPa, wood-decay species still grow at 30 kPa and some, including *Piptoporus betulinus*, still grow at 70 kPa. It appears that air equivalent to something more than 20% by volume of the wood is necessary for actual decay to take place. The existence of intact wooden galleys, submerged since Roman times, is adequate proof that completely water-logged wood does not decay. The point at which all the free water has disappeared, but the cell walls are still fully saturated, is known as the fibre saturation point. For most woods this is around 26–32% (c.0.3 $g g^{-1}$, equivalent to an $a_w$ of 0.97). Wood-decay fungi begin to grow at around this level and make optimum growth at about 40%. Tresner and Hayes (1971) tested just over 100 species of Basidiomycotina and found that 94 were unable to grow at an $a_w$ of 0.97 and below and only one species grew down to 0.94. Other evidence suggests that the lower limit for growth of wood-decay fungi is about 0.97, with the linear growth rate reduced to about half normal even at 0.989.

Worked wood that has been thoroughly air-seasoned contains 15–18% moisture, which is far too low to support any fungal growth. Requirements for such high moisture contents, and thus water activities, obviously contribute to the resistance of wood to decay. Dry rot, caused by *Serpula lacrimans*, is an exception. Wood with a moisture content as low as 20–24% becomes liable to attack by *S. lacrimans*. Furthermore, as a brown-rot fungus, it produces metabolic water during cellulose degradation which considerably raises the moisture content of the wood on which it is growing. Once established on a small pocket of damp wood it can continue to colonize dry wood in this way. The exact relationship between water activity and wood decay is difficult to obtain because all, like *S. lacrimans*, degrade the cellulose in the cell walls. The complete degradation of 1.0 g cellulose liberates 0.56 g metabolic water. This is sufficient to alter the $a_w$ of the wood significantly.

**Toxic Substances**

All these factors contribute to decay resistance but the principal sources of such resistance in wood are toxic substances deposited during the formation of the heartwood. These are synthesized in the senescing parenchyma cells and diffuse out into the walls of the adjacent xylem elements. The distribution of decay resistance within a tree has been correlated with both the distribution and the nature of these toxic substances. They have been studied most in Gymnosperms. Most are phenolics. They fall into four main chemical groups, terpenoids, tropolones, flavonoids and stilbenes (Fig. 6.4). Of these the thujaplicins (tropolones)

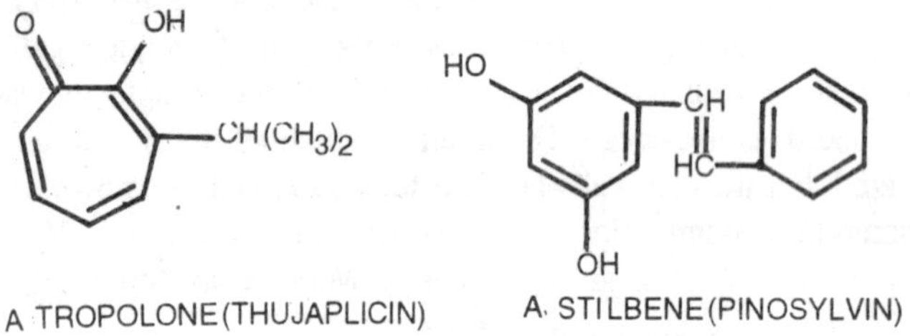


**Fig. 6.4.** Structure of two toxic chemicals extracted from gymnosperm wood.

are the most inhibitory. They all provide protection from decay for many years but with time they may become lost by leaching or become inactivated. In spite of their toxicity, it is well-known that several fungi are able to destroy the heartwood, even in living trees, and also timber impregnated with similar phenols, such as pentachlorophenol and 2, 4-dinitrophenol, which are used to protect less durable timbers and the sapwood of conifers. Such heartwood rotters are not insensitive to these toxins. They use their phenolases to oxidize them and polymerize the products to non-toxic melanins. Tannins are very common in the heartwood of Angiosperms and they play a similar role in decay resistance there. They inhibit fungal phenolases but decreased toxicity of the heartwood occurs with time by auto-oxidative polymerization of the tannins.

Sapwood is ordinarily very susceptible to decay but the resistance of different heartwoods is very variable. Trees with very resistant heartwoods include many oaks, cedars and the redwoods and those with non-resistant or only slightly resistant heartwoods include alders, beech, elms and poplars. A durable heartwood may be of survival value to the tree itself. Cedars live 2000 years or more whereas any of those in the slightly resistant category rarely live as long as 500 years.

## OTHER WOOD-INHABITING FUNGI

Other fungi which inhabit wood occur chiefly in the sapwood where they obtain their food supply from the contents of the dead xylem parenchyma cells. These are the so-called moulds and stain fungi.

Mould fungi are mainly conidial Ascomycotina. They discolour the wood by producing pigmented conidia on the surface. Their hyphae accumulate within the ray parenchyma cells but may also be present in the cell cavities of most of the surface xylem elements, spreading from cell to cell via pits. This causes shallow discolouration and surface straining of the wood. For instance, surface blue-stain occurs most frequently on sawn timbers and only any wood surface exposed to the rain. It is caused by the surface growth of common air-borne fungi, but especially *Cladosporium* spp., with dark brown hyphae and coloured conidia. Such staining is easily removed during planing treatments.

### Blue-Stain Fungi

Typical blue-staining is caused by pigmented hyphae that grow in the wood, whereas other stains, such as brown ones, are caused by chromogenic substances actually excreted by the hyphae into the wood. At least one fungal stain has been used commercially. The mycelium of the Ascomycete *Chlorosplenium aeruginascens* permeates the dead wood of oak and beech on the woodland floor and colours it a brilliant green. Such 'green oak' has been used for inlays and decoratively as Tunbridge Ware. The wood is unaltered in texture and resists decay.

Blue-stain fungi are common in coniferous sapwood, especially pines, but are also found in hardwoods. They are non-cellulolytic 'sugar' fungi, in that they utilize only the more readily assimilable carbon compounds, such as sugars and starches, which occur in the ray parenchyma cells of freshly killed wood. They do no structural damage to the wood as they move across it mainly through the pits. Blue-stain is thus not the first stage of a form of rot, but its occurrence does indicate that the wood has been kept moist and exposed to conditions favourable to the development of decay fungi. Although the structural properties of the wood are unaltered, blue staining of coniferous sapwood is responsible for large financial losses to the timber producer. The mere discolouration of the wood makes architects disinclined to use it and it is less acceptable to the manufacture of packing cases and paper. In vigorously growing pine trees, the moisture content of the sapwood is too high to permit growth of blue-stain fungi. The low oxygen tension again appears to be the major limiting factor in the growth of these fungi in wood with a very high moisture content. In nature they may colonize standing pine trees which have been killed either by root-rot caused by *Heterobasidion annosum*, or some other disease, or by suppression. They are much more common on felled pine logs and will soon appear on these if they are left on the forest floor for any length of time. However they are rapidly replaced by wood-decay fungi. Both death and felling cause the wood to dry out progressively and such wood will support the growth of blue-stain fungi unless the moisture content falls below about 27%.

Blue-stain fungi are Ascomycotina, mainly of the genus *Ceratocystis*, most of which have both perithecial and conidial states. The majority present their spores for dispersal in the form of stalked spore drops, the spores, in this case, being insect dispersed. The perithecia of *Ceratocystis* have a swollen base and a very long, slender neck, some often 1 mm or more in length (Fig. 6.5a). The ascospores are not violently discharged but the asci break down within the ascocarp and the ascospores are forced up the neck; they are extruded in a mucilaginous drop at the apex where they are held in place by a fringe of hair-like hyphae lining a pore. A variety of conidial states are produced. The *Graphium* state has a thick sheath of dark hyphae forming the stalk (Fig. 6.5b). The component hyphae branch at their tips and produce masses of sticky conidia, whereas, in the stalked spore drop of the *Leptographium* state, the stalk is a deeply pigmented and very wide single hyphae which branches profusely at the apex (Fig. 6.5).

Blue-staining is usually associated with attack by bark beetles. Species of *Hylastes, Myelophilus* and other introduce spores in making their brood chambers at the interface of sapwood and bark. The spores germinate and grow radially and longitudinally in the sapwood forming wedges of blue-stained timber and then produce their conidia projecting into the brood chambers. These adhere to the young beetles as they emerge and are dispersed to other logs as they in turn make brood chambers.

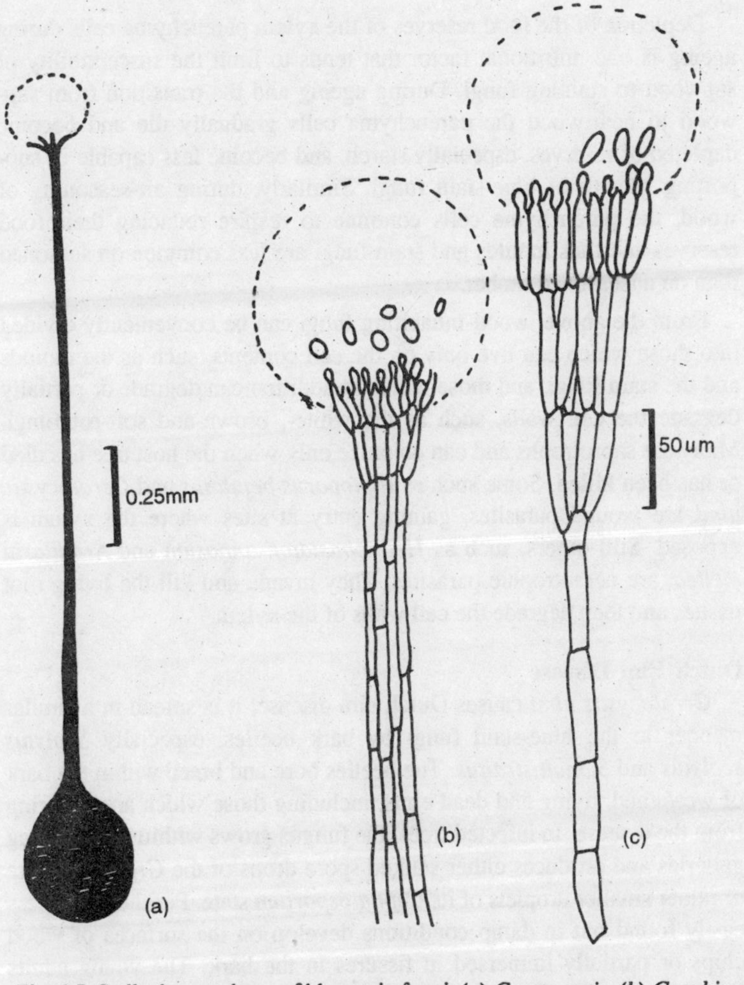


**Fig. 6.5.** Stalked spore drops of blue-stain fungi. (*a*) *Ceratocystis.* (*b*) *Graphium.* (*c*) *Leptographium.*

Blue-staining becomes a problem where felled pine logs are left in piles on the forest floor for 2–3 months before being removed to timber depots. There are a number of ways of treating the problem, such as the use of insecticides and fungicides, but the most successful method of control, widely used in Europe, is to remove the bark immediately on felling. This not only prevents beetle attack but assists rapid drying out to moisture contents below those which will support fungal growth.

Depletion of the food reserves of the xylem parenchyma cells during ageing is one nutritional factor that tends to limit the susceptibility of sapwood to staining fungi. During ageing and the transition from sapwood to heartwood the parenchyma cells gradually die and become depleted of reserves, especially starch, and become less capable of supporting growth of blue-stain fungi. Similarly, during air-seasoning of wood, the parenchyma cells continue to respire reducing their food reserves and thus moulds and stain fungi are less common on seasoned than on unseasoned timber.

From the above, wood-inhabiting fungi can be conveniently divided into those which can live only on the cell contents, such as the moulds and the stain fungi, and those which in addition can degrade or partially degrade the cell walls, such as the white-, brown-and soft-rot fungi. Many are saprotrophs and can colonize only when the host tree has died or has been killed. Some such as *Piptoporus betulinus* and *Ceratocystic ulmi* are wound parasites, gaining entry at sites where the xylem is exposed. Still others, such as *Heterobasidion annosum* and *Armillaria mellea*, are necrotrophic parasites. They invade and kill the living root tissues and then degrade the cell walls of the xylem.

**Dutch Elm Disease**

*Ceratocystis ulmi* causes Dutch elm disease; it is spread in a similar manner to the blue-stain fungi by bark beetles, especially *Scolytus scolytus* and *S. multistraitus*. The beetles bore and breed within the bark of weakened, dying and dead elms, including those which are suffering from the disease. In infected trees, the fungus grows within the breeding galleries and produces either stalked spore drops of the *Graphium* state or rather smaller droplets of its *Cephalosporium* state. Perithecia are less easily found but in damp conditions develop on the surfaces of wood chips or partially immersed in fissures in the bark. The young adults emerge in May to October and the sticky conidia may adhere to their bodies. They fly immediately to feed on young, healthy, elm twigs and in doing so may introduce the conidia into the xylem in wounds made as they feed. Beetles thus spread the fungus from branch to branch and tree to tree. The fungus enters the xylem and grows in a yeast-like form. It can be carried up in the xylem in the transpiration stream as such or as conidia. Infected trees soon show signs of wilting and yellowing or drying out of the foliage. Fungitoxins may be involved but part of the symptoms may be explained by the occlusion of the xylem of the current year's growth by gums and tyloses.

## ENVIRONMENTAL FACTORS AND THE DECOMPOSITION OF WOOD

The decomposition of wood under natural conditions is an exceedingly protracted process. Whereas leaves of the majority of northern temperate deciduous trees may decompose in one, two or three years, a tree trunk under the same conditions may take a decade or even two to do so. The low level of available nitrogen may be the overriding factor contributing to its slow rate of decay. The addition of organic nitrogen to wood blocks inoculated with various Basidiomycotina has been shown to increase their decay rate by over 60%. Other minerals, especially phosphorus and potassium, may also be limiting. The relatively high demand for such mineral nutrients combined with their relatively low availability places a limitation on the amount of fungal mycelium such a substrate can produce. Fluctuations, both diurnal and annual, in temperature and moisture content must also be important. However, in aseptic laboratory experiments decomposition of wood by a single species of decay fungus may be relatively rapid. The white-rot fungi *Lenzites betulina, Coriolus hirsutus* and *C. versicolor*, inoculated onto small blocks (20 $mm^3$) of birch wood kept in sterile moist soil at 22°C, caused more than a 75% loss in dry mass in three months. *Piptoporus betulinus* and a number of other brown-rot fungi caused mass losses of 50–70% over the same period. These are substantial losses, the more so when it is borne in mind that birch wood contains some 20% lignin which is not available to brown-rot fungi. These facts may be contrasted with the observation that *P. betulinus*, which had killed birch trees in East Anglia, was still producing basidiocarps on these at least five years after they had fallen. The time period from infection to falling was not known but it may have been at least another five years. Even if the variable temperature and moisture regimes are taken into account, the decomposition of one of the least durable woods is very much slower in nature than in laboratory tests.

Most wood-decay fungi are mesophiles in terms of their temperature requirements, although some come into the category of cold-tolerant ones. The optimum temperature for the growth of most lies between 25 and 30°C. *P. betulinus* has an optimum at 25°C and its growth falls off very rapidly above and ceases at 30°C. Its minimum temperature for growth, which it must often experience in the field, lies between 7 and 9°C. For many others, the minimum lies below freezing point but decay at such temperatures would be very slow. The geographical distribution of a number of species is related to their temperature requirements.

*Serpula lacrimans* has a low maximum of 25–26°C. It is absent from the tropics and other parts of the world with high summer temperatures.

In bulk wood, temperature is probably a more important variable than moisture content. With reference again to *P. betulinus*, it is able to decompose birch wood with a moisture content within the range 35–100% on a dry mass basis, although near maximum decomposition rates only occurred between 60 and 120% in laboratory experiments. Birch logs, stored outside in Central Sweden, had a moisture content of 85–91% on felling and after three years the moisture content was still 51–67%. Fluctuations did occur, with some drying in the summer and some water uptake in the winter, but over the whole period the moisture content was somewhere near the optimum for decay. However, if the bark peels off the position is quite different. In summer, rapid drying out may occur to moisture contents below those which will support growth and equally rapid soaking will occur in rain. A thick, highly suberized outer bark is not only a structural deterrent to fungi, but because of its high content of tannins, phenols and the like, also a chemical one. However, given this, if it remains intact after death of the tree, it helps to maintain a more equitable moisture regime within the wood and this will favour any decay fungi.

## HABITAT RELATIONS AND SPECIFICITY OF WOOD-INHABITING FUNGI

The habitats of wood-inhabiting fungi vary from minute twigs, small and large branches to the most massive of tree trunks and stumps and from minute rootlets to major roots and include such man-made habitats as fencing posts, house timbers, sawdust and chip piles. Any one of these substrates is particularly complex in more ways than one. The trunk of any one tree will have varying proportions of bark, sapwood and heartwood along its length. These proportions will differ from those in the trunk of another species. The wood from different tree species differs structurally as can be seen by contrasting ring-porous with diffuse-porous types. Further marked differences occur between softwoods and hardwoods. Over and above these differences as already indicated the composition of the lignin varies in different wood. This complexity and heterogeneity make it difficult to generalize about the decomposition process. A number of successional studies have been made on woody substrates but these have to be interpreted with caution. A succession can be defined as the appearance of different fungi in sequence on the same part of the substrate. The fact that one fungus appears on one part of a log at one time and another fungus on another part, even an adjacent part, at another time does not necessarily prove a succession. Their habitat niches may be quite different, one growing on the sapwood and one on the heartwood or the latter may be colonizing a part of the heartwood not colonized by the former. For example, basidiocarps of *Daedaleopsis confragosa* or of *Hypholoma fasciculare* may appear on a birch trunk which has, for a number of years, supported basidiocarps of *Piptoporus betulinus*, but the mycelia of these would almost certainly be growing on parts of the wood not colonized by *P. betulinus*. The latter is specific to birches and is a wound parasite gaining entry where a branch has been fractured. Infected trees are usually killed by the fungus and the trunks of these often break off remarkably cleanly and transversely at a height of about 3 m in high winds. The structural polysaccharides in the walls are rapidly and completely removed leaving a cellular framework of amorphous lignin which has insufficient tensile strength to withstand the bending strains incurred. The fungus then continues to grow and to produce its characteristic kidney-or hoof-shaped basidiocarps on the fallen and standing parts of the tree. As with many woody substrates, *P. betulinus* is the primary and sole colonizer. It may completely permeate the wood of the whole trunk and persist there, virtually in pure culture, for several years, by which time the wood is in a very late stage of decay and extremely friable. In such a state the wood is unlikely to be capable of supporting fungi such as *D. confragosa* and *H. fasciculare*. They would not succeed *P. betulinus* but would be growing on parts not colonized by it.

Wood-decay fungi exhibit all degrees of specificity. Considering the white-rot the brown-rot fungi as two groups, there are many more of the former than the latter. Those of the white-rot group primarily attack hardwoods and those of the brown-rot groups softwoods. Some of these may be restricted to a single host genus. *P. betulinus* is a good example. *Fistulina hepatica* which causes a serious decay of the heartwood of oaks is another. The cause of such marked specificity are obscure. Other fungi may be restricted to the wood of a relatively small number of trees, *Polyporus squamosus* is, like *P. betulinus*, a wound parasite, in this case of elm in particular but it is often found on other trees, such as ash and sycamore. It causes a white-rot of the heartwood and may persist for a number of years on fallen trees which it has killed or which have been wind-blown, as a consequence of the rot. On elm trunks *P. squamosus* is replaced, but only in a temporal sense, by a number of other wood-decay fungi. Two in particular, *Auricularia mesenterica* and *Pleurotus cornucopiae* are rarely found on other wood. Basidiocarps of the former soon appear on any felled elms and production of these continues for up to eight years. Spatially it utilizes the bark and surface layers of the sapwood so it does not succeed *P. squamosus*. Basidiocarps of *P. cornucopiae* appear on elm trunks only some 3–10 years after they have fallen. The fungus then persists until the wood is well-decayed. Its mycelium appears to be confined to the sapwood not utilized by *P. squamosus* and again it does not actually succeed the latter. Whereas *P. cornucopiae* is most common on fallen elm trunks, *Flammulina velutipes* is most often found on standing dead elms, especially those killed by *Ceratocystis ulmi* and which have lost their bark.

Still other fungi, such as *Coriolus versicolor* and *Stereum hirsutum*, are much less discriminating and grow on a wide range of hardwoods. The former is entirely saprotrophic and is one of the commonest fungi found on fallen twigs, branches, trunks and dead stumps of hardwoods where it produces the most rapid of white rots but, like *Stereum*, is confined to the sapwood. It can actually replace, and therefore succeed, other established and less aggressive white-rot fungi. *Xylaria hypoxylon* and *Daldinia concentrica*, two Ascomycotina, produce black lines, zone lines, in tbe sapwood of ash delimiting areas which they have colonized. The hyphae of *C. versicolor* will penetrate these and grow on to replace them. Other fungi show a preference for coniferous wood. *Heterobasidion annosum, Paxillus atrotomentosus* and *Tricholomopsis rutilans* are characteristic of conifer stumps, *Hirschoporus abietinus* and *Stereum sanguinolentum* of coniferous twigs and branches and *Auriscalpium vulgare* of pine cones. These are all Basidiomycotina but similar examples can be found in the Ascomycotina. For example, *Daldinia concentrica* is very common on ash but is occasionally found on other hosts, especially beech and birch. *D. vernicosa* occurs on gorse, especially bushes which have been burnt and subsequently weathered. *Ustulina deusta* causes a white-rot of lime and beech, whereas *Xylaria polymorpha* and *X. hypoxylon* are very common on a wide variety of dead hardwoods.

Each tree species thus may have, within limits, its own particular wood-decay fungi. A number of these may enter as necrotrophic parasites at wounds above ground or along roots below ground and then persist as active saprotrophs after death. They would thus have a competitive advantage over purely saprotrophic fungi in being established first. As parasites they may only be able to overcome the host resistance of one or a few species of trees. This might account for some of the specificity noted. It may well be that the different naturally occurring tannins, terpenoids, etc. present in the different heartwoods further help to determine specificity. Only fungi which can tolerate or degrade these are able to become established.

## ECOLOGICAL STUDIES ON DECAYING WOOD

Numerous ecological studies have been made of fungi colonizing specific woody substrates including wounded living tree trunks and fallen dead ones, tree trunks after insect attack, fire-killed trees, branches and slash on the ground, tree stumps, fence posts, beech cupules, etc. Changes with time in the fungal communities on these have been recorded and described, accurately or inaccurately, as successions. As might be expected, the sequences of fungi observed on these show considerable variation depending upon the species of wood, the type of substrate and, in addition, the environment in which decomposition is occurring. But fungi are not the only organisms found in decaying wood. A very wide variety of invertebrates and bacteria also occur and they, too, may play an important role in its decomposition.

Swift (1977) recognized three stages in the decay process—the pioneer colonization stage, the major decomposition stage and the incorporation stage, in which the products of decay are incorporated into the soil.

### Pioneer Colonization Stage

Patterns of colonization may vary. In some cases, the Basidiomycotina which are going to dominate the decomposition stage are the primary and sole colonizers. In others, their colonization is preceded or accompanied by a variety of decay or non-decay fungi or bacteria. This may be illustrated with some specific examples. *Heterobasidion annosum* is a white-rot fungus causing butt-and root-rot of conifers. It may colonize via roots or the surfaces of freshly cut stumps. Infection of a healthy living root almost invariably occurs as a result of mycelial transfer from another infected root coming into contact with it. From the root the fungus grows up to the base of the stem and colonizes and kills the cambium, thus effectively girdling and so killing the tree. It then progressively rots the roots and the stem base. Rapid desiccation of the wood after death usually prevents extensive spread up the stem. In this case it is the sole colonizer. Alternatively it may colonize the surfaces of freshly cut stumps via its air-borne basidiospores. These stump surfaces are highly selective substrates and are initially colonized by a relatively small number but, nevertheless, a variety of fungi. These include, in addition to *H. annosum*, non-cellulolytic blue-stain fungi utilizing the contents of the parenchyma cells, cellulolytic fungi, such as *Phialophora* and *Trichoderma* spp., utilizing cell contents and any easily accessible cellulose, and other wood-decay fungi such as the white-rot fungus *Peniophora gigantea*. This is a much more competitive situation; whether or not it emerges as the major decomposer will depend upon a multiplicity of factors, including its ability to compete with these for the more readily available nutrients which are necessary if it is to become established. Similar patterns of colonization can be seen in the initiation of decay in trunks following wounding, such as by the branches breaking off in high winds. Again, in some cases the only fungi to colonize are the wood-decay Basidiomycotina which later become the dominant decomposers. This applies to most species of *Stereum*. They invade only freshly exposed tissues and are inhibited by the presence of other pioneer micro-fungi and bacteria. In other cases, such as with *Phellinus igniarius* invading wounds on poplars and other hardwoods, prior colonization by bacteria and micro-fungi such as the stain, mould and soft-rot fungi generally occurs and may even be a prerequisite if it is to attack and cause a progressive rot of the heartwood.

Insects, especially members of the Ipidae and Scolytidae, may attack living trees and introduce bacteria, yeasts, blue-stain or ambrosia fungi below the protective bark. The combined activities of the insects and the fungi may weaken or kill the tree. The wood-decay Basidiomycotina then follow. The attack of *Scolytus scolytus* on elms introducing *Ceratocytsis ulmi*, followed by *Flammulina velutipes*, is a case in point.

**Decomposition Phase**

The decomposition phase is dominated by the white-and brown-rot fungi but wood-boring beetles (Coleoptera) and wood-eating termites (Isoptera) may also contribute to decay. In many woody substrates only one fungus may be involved in the decomposition phase; examples of *Piptoporus betulinus* on birch, *Heterobasidion annosum* on pines and *Coriolus versicolor* on hardwoods in general, have already been given. In others, a number, but usually a very limited number, of fungi are involved. Each of these occupies discrete volumes of wood which are often clearly demarcated from each other by distinct dark zone lines. These colonies may intricately interlock but their mycelia do not intermix. They remain isolated by zone lines into virtually pure cultures. This balanced state may persist for a number of years, but, depending upon the relative competitive ability of adjacent mycelia, there may be eventually some replacement of one fungus by another, or aggressive saprotrophs, such as *Hypholoma fasciculare, Phallus impudicus* and *Phlebia merismoides*, may colonize from the surrounding litter and replace them. For example, both *C. versicolor* and *Stereum hirsutum* are susceptible to replacement by any of these three in hardwood trunks and branches and *Heterobasidium annosum* by *Peniophora gigantea* in pine stumps. *H. annosum* is particularly sensitive to hyphal interference caused by the latter and this may be one factor involved in its replacement.

**The Role of Animals in Degradation**

Many wood-boring beetles and their larvae and termites are wood feeders depending upon microbial symbionts in their guts to semi-digest the wood. Some feed on sound wood, others on decaying and well-rotted wood. In the latter case, the fungi growing in the wood may be an important component of their food. Many termites are polyphagous. When *Kalotermes flavicallis* is fed on wood, it decomposes 94–95% of the cellulose, 60–70% of the hemicelluloses and 3–4% of the lignin. Thus large populations of these wood-boring beetles would almost certainly contribute substantially to wood decay. Another important aspect of their degradative activity is the comminution of the wood as they attack it.

As the white-and brown-rot fungi exploit the wood, it softens and becomes friable and as such is more attractive to animals as a food source, as somewhere to live and as a breeding ground. Their access to it, if it is bulky, may be dependent upon the prior activity of wood-boring animals, such as the beetles. Their bore holes afford ports of entry for a very great variety of animals generally common in litter and soil. These include micro-arthropods, such as Acari and Collembola, and macro-arthropods, such as Diptera and Isopoda, as well as Oligochaetes, such as Enchytraeid and Lumbricid worms. Many of these, such as the Mycetophilid dipterous larvae feed mainly on the mycelia of the fungi. They all accelerate the process of comminution and carry spores from the surrounding litter and soil into the wood and thus inoculate it with common soil fungi. Many Zygomycete Mucorales appear for the first time on the decaying wood, along with a variety of conidial fungi including species of *Penicillium, Scytalidium* and *Trichoderma*. These now have a quite wide choice of substrates on which to grow. They may utilize the partly degraded wood, the dead hyphae of the wood-decay fungi, dead fauna or their faecal remains. Some may live as commensals sharing the hydrolytic products of the enzyme systems of the major decomposers. This is the incorporation stage. As the wood becomes more extensively decayed, the activity of the wood-decay fungi may decline and they are eventually replaced by such soil-inhabiting fungi. With time, the wood disintegrates and as it does so it is incorporated into the soil.

### Wood Decay and the Cycling of Mineral Nutrients

Wood decay is important in regulating the cycling of mineral nutrients in the woodland ecosystem and contributes to the process of soil development there. Over the period of fungal decay, virtually all the important minerals, but in particular nitrogen and phosphorus, become immobilized in an organic form in the fungal hyphae and their reproductive structures, such as basidiocarps. Although the mineral content is low per unit volume, the sheer volume of decomposing wood means that it forms a very substantial part of the total minerals in the woodland ecosystem. Comminution of the decaying wood by the animal invaders leads to a release of some minerals. The small particulate form of the frass or faecal materials means that they are more effectively leached. As with the fungi, the animals themselves act as a further reservoir of plant nutrients in a considerably more concentrated form than in the wood itself. Their wanderings, after feeding, lead to some redistribution of minerals, but by far the more major redistribution and export from the wood occurs when adult stages emerge from the broods reared in and on the decaying wood. But this is essentially only a redistribution. The adults eventually die and their tissues are mineralized elsewhere in the ecosystem, while any fungal remains are mineralized *in situ.*

## DECOMPOSITION OF LIGNIN AND HUMUS IN THE SOIL

The white-and brown-rot fungi are, more often than not, associated with relatively large masses of wood, such as the dead tree trunk and decaying stump. This may be because only these contain enough energy resources for these fungi to amass sufficient to produce their relatively massive and conspicuous basidiocarps. Vast quantities of lignin are incorporated into the soil in the vascular network of the leaves, fine rootlets and so on. These are very different substrates for fungi and other micro-organisms and they are in a vastly different environment. The substrate is richer in terms of associated readily available carbon and nitrogen sources, as tissues other than the highly lignified xylem are present in relatively larger proportions than in bulk wood. It is also presented to a much more varied population of micro-organisms. As such it would support a more diverse micro-flora and any lignin decomposers would be competing for, not necessarily lignin, but other more generally assimilable components, which are necessary for establishment, and would also be exposed to antagonism by others. This situation is markedly different from the decaying tree trunk with its one or few decomposer fungi in isolation. Further large quantities of lignin may be introduced into the soil in the form of organic residues from wood decay, especially from brown rotted wood. The process of lignin degradation in the soil may be quite different from that occurring in a tree trunk. There is very little direct evidence that any Basidiomycotina degrade such lignin in the soil. This may be because we are ignorant of the facts. In studies on soil fungi, Basidiomycotina are only rarely recorded. They tend to be slower growing and so are easily overgrown on most widely used culture media. They are often very sensitive to antagonism by others and so suppressed. Most do not produce spores or possess any other readily recognizable feature so could easily be overlooked. Nevertheless, they may be equally important as lignin decomposers in the soil itself as they are in the litter and decaying wood. The number of soil-inhabiting mirco-organisms which have been reported as being able to utilize lignin is very small. These include a few aerobic, Gram-positive, non-sporing, rod-shaped bacteria, in the genera *Bacillus* and *Flavobacterium*, and a few conidial fungi, in genera such as *Humicola* and *Phialophora*. The evidence for their ability to utilize lignin has again been obtained from the use of extracted lignin and lignin model compounds. These have been used incorporated in Kaolin pellets to enrich soil and fungi subsequently isolated from them and tested for their ability to utilize such compounds as vanillic acid and syringaldehyde. The ability to grow on and utilize these should not be taken as an ability to utilize lignin itself just as the ability to utilize carboxymethyl cellulose is not taken as an ability to utilize native cellulose. They are partial degradation products and these fungi should be regarded as occupying a similar niche with regard to lignin as secondary sugar fungi do to cellulose. The latter do not possess the whole enzyme system necessary to hydrolyse cellulose (Fig. 6.6). They lack the $C_1$ component but possess the $C_x$ component and β-glucosidase so that they can utilize the hydrolytic products. Some also

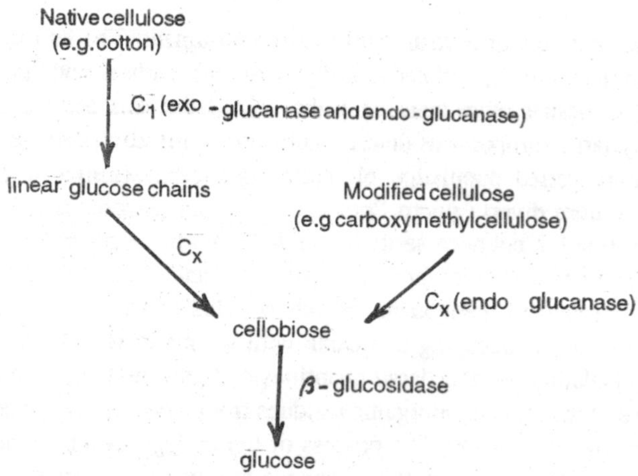


**Fig. 6.6.** A schema for the degradation of cellulose.

lack the $C_x$ component as well. Similarly with lignin, the fact that a fungus lacks one component of the multi-enzyme system does not necessarily debar it from participating in lignin degradation. Fungi may co-operate, sometimes synergistically, in the degradation of both cellulose and lignin. For example, it has been shown in experimental systems that a mixture of the $C_1$ component from one fungus and the $C_x$ component of another is as efficient at cellulolysis as when both components are derived from the same fungus. It has also been shown using lignin preparations in culture tests that, in many cases, when two wood-decay fungi are grown together in mixed culture, degradation is more pronounced than when both fungi are grown apart.

Lignin in the soil decomposes very slowly and its degradation there is more of a joint effort. There may well be a pooling of enzymes from a variety of fungi and perhaps bacteria and actinomycetes—some enzymes capable of cleaving bonds between monomers, others of demethylation and still others of side chain oxidation and so on until the final products enter the respiratory pathways of one organism or another.

## The Nature of Humus

With this breakdown there is a gradual accumulation of dark, amorphous, organic humus. The chemistry of humus has by no means been fully elucidated. It is very heterogeneous and can be separated into a number of molecular categories using extraction techniques. It forms a very dark solution in dilute NaOH and a black precipitate called humin. Acidification of the solution with HCl to between pH 1–2 precipitates out a fraction called humic acid, leaving fulvic acid. The humic acid fraction is the major molecular category and forms from between 50 to 80% of the soil humus. It usually contains about 5% nitrogen, mainly in the form of bound amino acids but also in amino sugars and heterocyclic purine or pyrimidine derivatives. The most favoured idea is that humic acid has a heterogeneous aromatic core with carbohydrates, peptides and proteins, phenolics and metals, attached peripherally. In some soils, especially under woodlands, the humic acid may originate from lignin residues, possibly the end products of the brown-rot fungi, which have been considerably modified by microbial action. Syringic and vanillic residues can often be detected in the degradation products of humic acid and the distribution of these residues is consistent with the composition of the lignin found in the vegetation above. Work with tracers has shown that as much as one third of the humic acid in the soil is derived from lignin and only about one twentieth from cellulose. Reductive cleavage of most humic acid fractions shows that they also contain units based on phloroglucinol. This suggests that seed plant flavonoids also contribute to humus. Flavonoids are phenolics with two aromatic rings and include pigments, such as anthocyanins. The phenolics are degraded to simple phenols which become polymerized into the humic acid fraction. But humus is not solely a product of degradation of the more resistant parts of seed plants. It is also in substantial part a product of microbial synthesis. When $^{14}C$ labelled glucose is added to the soil, 40–80% of the carbon is lost as carbon dioxide within a few days but, even after two years, about 5–10% is still present in the soil humus. Intracellular transformation of carbohydrates and other simple organic substances occurs to produce phenols, quinones and other aromatic substances. These are oxidatively polymerized and combined with peptides and other cell constituents to form humic-like pigments, melanins, inside or outside the cell. These serve several functions. They may be deposited in the walls of hyphae, spores or ascocarps, to protect against excessive ultra-violet light or as a water-proofing to prevent water loss. Eventually, on the death of these structures and with time, they become variously transformed and incorporated into the humus fraction. The existence of significant amounts of amino sugars and non-protein amino acids, such as diaminopimelic acid, also suggests that residues of bacterial cell walls may form part of the humus.

**Turnover of Humus in Soil**

Humus is extremely resistant to microbial degradation but nevertheless there is a very slow turnover with the rate depending upon the soil type. A sample from a chernozem soil from the USA was $^{14}C$ dated as $990 \pm 60$ y old.

In other soils, humus is less stable e.g. humus from a coniferous forest soil in Sweden was dated as $370 \pm 10$ y. A number of fungi have been found to decompose humic acid in laboratory tests. Humic acid was extracted from a Canadian soil. It contained 26% of the total soil carbon and was dated as $785 \pm 50$ y old. It was supplied as the sole carbon and nitrogen source as a 0.2% solution to a number of microorganisms, isolated by direct plating of the soil onto humic acid containing media. Four bacteria, in the genera *Bacillus* and *Pseudomonas*, and two conidial fungi, *Penicillium frequentans* and *Aspergillus versicolor*, could utilize the humic acid as a sole carbon and nitrogen source but no actinomycetes could. *P. frequentans* made the best growth and it appeared to utilize the humic acid by initially reducing carboxylic groups to aldehydes and then alcohols. Salicylaldehyde and salicyl alcohol appeared in culture filtrates. A number of Basidiomycotina, including *Coriolus versicolor, Hypholoma fasciculare* and *Trametes suaveolens*, all active white-rot fungi, can also utilize humic acid and this ability is always associated with the reduction of carboxylic acids to alcohols. This suggests that one of the first steps in the degradative process is an aerobic reductive one. Subsequent steps have as yet to be elucidated. Very little can be concluded from such studies about the process of degradation in soils. Resistance to degradation may not be so much that it is not susceptible to microbial enzymes but that its multi-dimensional complex structure physically restricts the access of such enzymes.

# 7

# CHEMICAL TREATMENT OF FUNGI

This chapter will deal with the intrinsic properties of compounds that make them fungitoxic.

As is the case of almost any discussion of chemical properties relating to biological action, there will be considerable simplification. Killing of living cells by a chemical compound is not a simple process, but rather the total observable response to a single or multiple chain reaction. Whether or not a specific property of a compound will result in the death of a specific cell or tissue very often depends on an exquisite interaction between the physical-chemical properties of the compound and those of the cell affected. For example, fluoroacetic acid is produced in the poisonous plant *Dichapetalum* without harm to the plant. Peters (1952) has found that when mammals ingest the plant, the fluoroacetate is mistaken for acetate in the Krebs cycle and fluoroacetate is produced which inhibits aconitase. This causes citrate to accumulate. Peters feels that the excess citrate is the toxicant. Here is a delicate change with a profound effect. Fluoroacetate is not a direct poison, but may be changed by mammalian tissue into another compound, fluorocitrate, which causes the accumulation of the toxicant. It would indeed be an oversimplification to ascribe to one property of fluoroacetate the ability to kill directly.

What must a compound be able to do in order to poison a cell? Simply stated, a poisonous compound must be able to get to a site of action, resist changes in transit, and exert a deleterious effect when it arrives. Requirements for poisonous molecules will, therefore, differ according to where they exert their effect. If they act outside the cell or at its surface, they will have no need to penetrate the intricate cellular membranes. In this case, the requirements for toxicants would be much simpler and less specific than for compounds which must penetrate.

Most organic molecules enter living cells most efficiently in the nonionized state. Organic toxicants, therefore, are usually most toxic in their nonionized form.

Are molecules always most toxic in their nonionized form? Albert (1951) and his group have found that the bactericidal effectiveness of the acridines and related series depends on their ability to ionize. Also, the addition of an aliphatic side chain, which usually aids permeation and attendant toxicity, actually reduces the bacteritoxicity of these compounds. Albert has deduced that acridine must act at the surface of the bacterial cells.

The 8-quinolinol derivatives, on the other hand, must be nonionized for maximum effect. If, as in the case of 8-quinolinols, the toxic molecule must enter the turbulent laboratory of the living cell, the molecule must resist destruction before it reaches its site of action. Rich and have demonstrated correlation between the detoxifying enzymes present in fungi and the resistance of these fungi to poisoning.

What does this mean in terms of comparative reactivity? A toxic molecule which is highly reactive and relatively unstable would be most likely to act in the outer portions of the cell. Great reactivity might serve to destroy and detoxify such a molecule before it moved very deeply into the cell. This property may, in fact, destroy a compound in the ambient fluid before it reaches the cell. For example, the presence of cysteine in the ambient fluid destroys captan before it can poison yeast cells. Lukens and Rich find that if yeast cells are treated with cobalt and then washed free of excess cobalt, they are more susceptible to captan than cells not treated with cobalt. Assuming that cobalt ties up surface sulfhydryl groups, the cobalt pretreatment presumably prevents the partial breakdown of captan on its way to its site of action within the cell.

A more stable, relatively unreactive molecule would have a greater chance of penetrating unchanged to a site of action deeper within the cell. It will be seen that compounds considered chemically inert may be toxic.

Not only the relative site of the action, but also the mode of action dictates requirements for a toxic molecule. For example, a competitive antimetabolite has more and subtler requirements than those of a nonspecific poison, such as phenol. Sulfanilamide must not only be able to penetrate to a site of action and be resistant to detoxication, but in addition its molecular configuration must very closely approximate the metabolite *p*-aminobenzoic acid with which it competes within the cell. The distances between atoms in the molecule and their spatial relations are critical to the action of sulfanilamide.

Much has been said of the solubility of toxicants in relation to their toxicity. Does a compound have to be water-soluble to be toxic? How else can it be reactive in the water-flooded interior of the cell? Actually, such materials as copper oxide, sulfur, and dichlone, highly insoluble in water, are toxic. Very often water solubility is inversely related to toxicity as in the analogues of metallic 8-quinolinolates. This does not imply that good solubility in water precludes toxicity. The organic mercury salts, highly water-soluble, are excellent fungicides. Usually those toxicants which are water-soluble, are very reactive compounds exerting general nonspecific toxicity.

Why this wide variation in the relation between toxicity and water solubility? As the sites of action may be in polar or nonpolar milieu, or in a medium with both polar and nonpolar groups, e.g., proteins, toxic compounds must and do vary in their solubilities depending on where they must go to become toxic.

Most often in the case of organic toxicants, of primary importance is not water solubility or lipid (nonpolar) solubility, but rather the partitioning of the compound between the two phases. Rich and Horsfall (1952) state the case for the importance of this relationship in the action of fungicides. Working with a series of nitrosopyrazole analogues, they showed that increasing the oil-water partition coefficient from 0.48 to 6.66 increases fungitoxicity ten thousandfold with no appreciable change in the dosage response slopes of the analogues. They interpret this to mean that the change in the direction of high oil-water partition has greatly increased the penetration of the compound into the cell without altering the mode of action.

This solubility balance may be so critical that it may decide whether a compound will be toxic to one plant species and not to another. Wellman and McCallan (1946) working with 2-substituted imidazoline derivatives found that the tridecyl derivative was fourteen times more phytotoxic than the heptadecyl derivative. The heptadecyl homologue, however, was twice as toxic to fungi as the tridecyl homologue.

A more complicated relationship between toxicity and comparative solubilities exists in the 8-quinolinol derivatives. Albert and Hampton (1954) and Albert et. al. (1954) showed that the antibacterial action of oxine derivatives is directly and closely correlated with oil-water partition. Albert et. al. (1953), with bacteria, and Block (1956) and Sijpesteijn et. al. (1957), with fungi, have demonstrated by various indirect biological methods that the toxic species of oxine and its derivative is the water-soluble half chelate. When the system is flooded with oxine well in excess of that needed for full chelation, there is no toxicity. Excess oxine would maintain the full chelate within the cell, preventing it from going to the half chelate. Thus, high oil-water partition is essential for movement of the molecule to a site of action, but, once there, the water soluble half chelate must be present for toxicity. This does not imply that water solubility per se is essential to the toxicity of the half chelate but merely that high oil-water partition is essential to movement, and once inside the cell this condition is not essential to toxicity.

The toxicity of another group of compounds appears to be completely correlated with oil-water partition, with no evidence that any other property is involved. These are the so-called inert, nonspecific toxicants which follow the Ferguson effect, to be discussed later.

The properties aiding direct permeation into a cell may not be the only ones of importance in moving a damaging number of toxic molecules to the site of action. Sussman (1954) showed that dormant ascospores of *Neurospora* adsorb ethylenediamine tetraacetic acid to their surface. The adsorption takes place even if the spores are first killed by boiling. If the ethylenediamine tetraacetic acid is removed from the spores before their dormancy is broken, there is no toxicity, but ethylenediamine tetraacetic acid is toxic of present when germination processes begin. Sussman interprets this to mean that the ethylenediamine tetraacetic acid cannot enter the dormant spore. Not until the permeability properties of the spores change by the onset of germination can the toxic molecules enter. Lowry et. al. (1957) presented evidence that metallic cations such as $Ag^{+}$, $Cu^{++}$, and $UO_2^{++}$, follow this same course. Here, then, is an example of adsorptive forces concentrating a toxicant at the surface of a cell, ready to enter when the permeability properties of the cell change.

The work of Sussman and his colleagues should not be taken to mean that metallic ions enter only in the fashion they describe. Miller et. al. (1953) have shown that fungus spores take up silver and cerium extremely rapidly and greatly concentrate the toxicants in so doing. With silver, for example, as high as 2124 p.p.m. of spore weight may be taken up by *Monilinia fructicola* spores in the first 30 seconds of exposure to a maximum possible dose of 4160 p.p.m. Cerium entered *Aspergillus niger* spores at the rate of 7625 p.p.m. of spore weight in the first 30 seconds of exposure to a maximum possible dose of 10,000 p.p.m. Miller et. al. (1953) point out that if doses of this magnitude were the result of simple adsorption, then the spore surface would have to be covered with multimolecular layers of toxicant. This does not appear likely.

Owens and Miller (1957) later studied the distribution of cerium and silver toxicants in spores. These workers disintegrated toxicant-treated spores by sonic vibration, then fractionated the spore fragments by centrifugation. They found that only 4% of the total amount of cerium or silver taken up was associated with the cell walls. Owens and Miller feel that adsorption of cerium or silver on spore walls or plasma membranes plays little or no role in the movement of these toxic ions into the conidia of their two test organisms, *Neurospora sitophila* and *Aspergillus niger.*

The following two sections discuss specific inorganic and organic toxicants in detail. The last section will deal with fashioning fungicides and with the future of fungicides.

## I. DESTRUCTIVE PROPERTIES OF FUNGITOXICANTS

Very few fungitoxic mechanisms are fully understood. Possession of one of the properties discussed below does not necessarily make the molecule toxic, nor does it explain why the molecule is toxic. Rather, this is a discussion of chemical and physical properties by which molecules may disrupt essential life processes.

### A. Direct Chemical Interactions

Toxicants may undergo direct chemical interactions with vital constituents of the cell. For example, irreversible addition or substitution reactions with important enzymes or substrates would effectively poison a cell. Even reversible reactions will inhibit manifestations of life, and eventually cause death if a continuing supply of the toxicant is fed to the cell. These reversible reactions, inhibiting only as long as the toxicant is present, exemplify fungistasis or bacteriostasis.

Whether or not a toxic reaction is reversible, will depend largely on the type of bonds formed between toxicant and vital component. Albert (1951) summarizes bonds of importance in biology and gives their usual bond strengths. The *van der Waals' bond* is a weak bond (about 0.5 kilocalories) which may occur when any two atoms or two different molecules approach each other. Atoms must be able to approach each other very closely if they are to be firmly bound by van der Waals' bonds. Although a single one of these bonds is weak, there may be many of them between molecules of toxicants and vital constituents. The formation of many van der waals' bonds is more likely to occur between molecules which can fit close together because of their molecular configurations.

*The hydrogen bond*, with a strength of 2 to 5 kilocalories, is fairly common in biology. In biological systems the hydrogen bond donors are usually the groups =N⁺—H, =N—H, or —O—H. The hydrogen bond accepts oxygen or nitrogen atoms, or sometimes chloride ions.

The hydrogen bond is apparently made by resonance between two structures where hydrogen is attached to one or the other of the atoms which it joins. There is supposedly no actual oscillation, but the distribution of electrons is intermediate between the two forms [A——HB] and [AH——B] and is represented by [A——B]. The hydrogen bond is stable, because the energy of the mesomeric state is less than that of either of the other structures.

*Ionic bonds*, with a usual strength of about 5 kilocalories, form between oppositely charged ions. This is the bond of salt formation.

By far the strongest is the *covalent bond* with bond strengths of from 40 to 100 kilocalories. This bond, in which combining atoms each contribute one shared electron, makes for essentially irreversible combinations between toxic molecules and vital components. Such reactions as alkylation, acylation, or halogenation are most likely to cause irreversible poisoning. For example, Burchfield and Storrs (1956) suggest that 2, 4-dichloro-6 (*o*-chloroanilino)-*s*-triazine may be so effective as a fungitoxicant because it alkylates amino acids.

Factors other than bond strength are also important in toxicity by direct chemical interaction. One of these factors is reaction rate. Burchfield and Storrs (1957) studied the importance of reaction rate in the activity of toxicants. Specifically, they investigated the kinetics of reaction *in vitro* between toxicants and vital constituents of cells. Although two toxicants may possibly react with the same group of a vital constituent, the ease and speed with which the reaction occurs may well decide whether the reaction is important to the toxicity of one compound and not the other. Some of the factors discussed were energies of activation, frequency factors, degree of solvation of a reactive halogen, and the degree of dissociation of the substrate. 1-Fluoro-2, 4-dinitrobenzene was found to be a more highly specific reagent for sulfhydryl groups than is 2, 4-dichloro-6 (*o*-chloroanilino)-*s*-traizine. Conversely, the *s*-traizine is highly reactive with organic amines and phenols, while the 1-fluoro-2, 4-dinitrobenzene is not. This latter difference is apparently related to the low dissociation constants of organic amines and phenols important to the growth of the organism.

Lukens and Sisler (1958) suggest that the toxicity of captan is related to double decomposition reaction of captan with sulfhydryl containing components of the cell. Here again, whether or not a compound will react in a certain way is not enough to make it a toxicant. The rate of reaction must be sufficiently fast to be effective. For example, if captan is toxic because of the reaction of —S—C(Cl)$_3$ with essential sulfhydryl groups, then the rest of the molecule may decide the toxicity of the compound as a whole. The phthalimide moiety of captan has carbonyl groups on each side of the N. These electron attracting groups weaken the N—S bond. If these electronegative groups were not present, or not in conjugation with the N, the resulting N—S linkage would be much more stable. The reaction rates would be slowed down and toxicity would be lessened. For example, our data show that *N*-trichloromethylthiocarbazole is much more stable than captan and also practically nonfungitoxic.

The heavy metals, such as silver and mercury, combine readily and firmly with sulfhydryl groups. Without question, this property contributes to their toxicity. Certain well-known specific reagents for sulfhydryl groups, such as iodoacetamide, are also toxic.

Not only are metals toxic because they may combine readily and firmly with cellular constituents, but also other molecules may be toxic because they combine with metals which are necessary to life. This latter mechanism has been proposed for chelating agents. Binding metals is presumably the action of azide and cyanide.

Bisulfite, toxic to many organisms, is known to combine with pyruvate. This combination would inhibit the decarboxylation of pyruvate, essential to organic acid metabolism.

The α, β-unsaturated ketones, including quinones, form addition products with thiols and amino compounds. This property has been suggested as a basis for toxicity.

There are, then, a number of direct reactions between foreign molecules and cellular components which may lead to toxicity.

### B. Effects on Enzyme Systems

One of the most popular areas of research on toxicants has been the study of their effects on enzyme systems. The primary reason for this has been the need to explain the profound effects produced by the application of extremely low dosages of toxicants. A most reasonable explanation is the assumption that the toxicants hit the organisms in a vital link in their metabolic chain. The many enzymes needed by organisms to change raw food materials into energy and living substance loom large as potential targets. These organic catalysts are not only themselves susceptible to inactivation, but also the processes they perform may be so altered as to result in toxicity. When these effects are described, therefore, they are presented in terms of effects on enzyme systems.

1. *Enzyme Inhibition*. The three principal kinds of inhibition of enzyme systems are termed competitive, noncompetitive, and uncompetitive.

Competitive inhibitors presumably poison by so closely resembling vital substrate that they are taken up by the enzymes which normally work on these substrates. The inhibitor molecules, however, differ sufficiently from the normal substrate molecules so that they cannot be changed by the enzymes into nutrilites needed by the cell to survive. The insertion of these ill-fitting keys into the active sites on the enzyme exclude the required substrate. The competition between the essential molecules and the imposter molecules for enzyme sites is in proportion to the ratio of the two competing molecules in the system. In other words, the degree of enzyme inhibition is proportional to the ratio of substrate to inhibitor. The more substrate, the more inhibitor required for the same degree of inhibition, and vice versa.

The classic Lineweaver-Burk (1934) plots of enzyme kinetics compare the reciprocal of substrate concentration *S* to the reciprocal of the observed reaction rate *v*. Figure 7.1 represents a typical plot for a case of competitive inhibition.

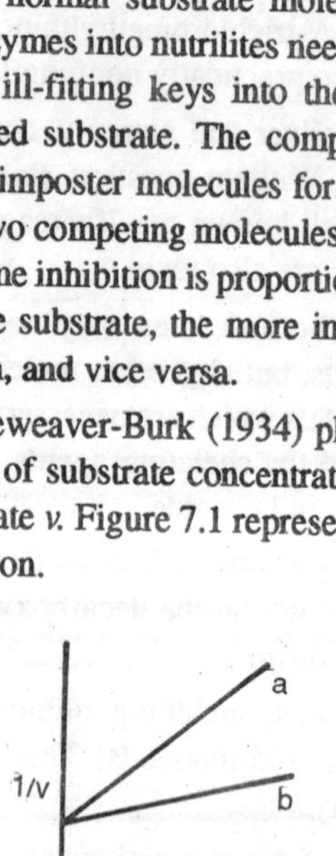


**Fig. 7.1.** Lineweaver-Burk plot for competitive inhibition.

The slope *V* of the inhibited line (a) is much steeper than the slope of the noninhibited line (b). These intercepts with the ordinate, however, are identical.

Equation 1 describes the steady state for competitive inhibition. At steady state $V$ = maximum velocity when the enzyme-substrate combination is maximum.

$$v = \frac{V}{I + K_S\,[I + (I)/K_i]/S} \qquad \text{(Eq. 1)}$$

$K_S$ = Michaelis constant, which is the substrate concentration at which half maximum velocity is reached.

$I$ = Concentration of inhibitor.

$S$ = Concentration of substrate.

$K_i$ = Constant for dissociation of enzyme-inhibitor combination into enzyme and inhibitor.

As $S$ is increased it affects both $K_S$ and $V$. When $S$ is sufficiently large, inhibition is completely overcome, and the intercept at sufficient value of $S$ becomes identical with that of no inhibitor. The slopes of the two systems, one with and the other without inhibitor, differ by the function $[I + (i)/K_i]$.

Noncompetitive inhibition applies to the case where the inhibitor combines with the enzyme at a point other than the site at which the substrate attaches. There are two other requirements for this type of inhibition: (a) the inhibitor exerts a total inhibitory affect; and (b) the binding of inhibitor to enzyme is independent of substrate. Under these conditions, the Lineweaver-Burk plot shows lines differing in both slope and intercept (Fig. 7.2).

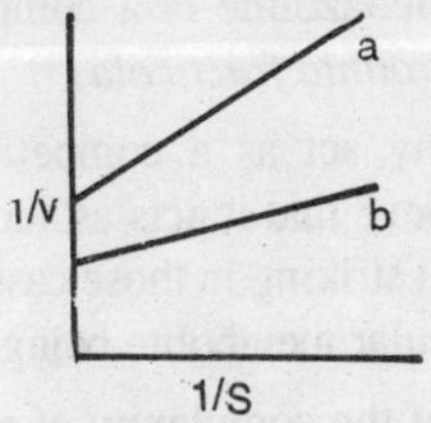


Fig. 7.2. Lineweaver-Burk plot for noncompetitive inhibition.

Equation 2 describes such a system.

$$v = \frac{V}{[I + K_S/S]\,[I + (I)/K_i]} \qquad \text{(Eq. 2)}$$

Here, as in the preceding equation, the slope of $V$ is still influenced by the inhibitor function, so that the slope of the inhibited system varies from the slope of the noninhibited system by $[I + (I)/K_i]$. It differs, however, by not having the inhibitor function modified by substrate concentration. Hence, at the intercept, the lines are separated by $[I + (I)/K_i]$. In other words, the reaction rate of a system noncompetitively inhibited can never attain the rate of the noninhibited system, regardless of the amount of substrate added.

Uncompetitive inhibition requires that the inhibitor combine only with the enzyme-substrate complex. In these circumstances, the Lineweaver-Burk plot shows parallel slopes for the two systems, and of necessity, different intercepts. Equation 3 describes uncompetitive inhibition.

$$v = \frac{V/[I + (I)/K_i]}{I + K_S\,[I + (I)/K_i]/S} \qquad \text{(Eq. 3)}$$

Other special cases include inhibition by a partial inhibitor, and inhibition by one of the products of the normal substrate with enzyme.

What does all this mean in terms of *in vivo* toxic effects? The great interest in competitive inhibitors as a toxic mechanism stems from the classic research of Woods (1940), who demonstrated that sulfanilamide is a competitive inhibitor of *p*-aminobenzoic acid. Not until after Woods' report was *p*-aminobenzoic acid found to be an essential metabolite. Since that time, whenever a toxicant is found which bears a structural resemblance to a vital substrate, competitive inhibition is invoked.

Actually, proven cases of competitive inhibition as a fungitoxic mechanism are not too common. West and Wolf (1955) present strong evidence that 2-heptadecyl-2-imidazoline is a competitive antagonist for guanine and xanthine in *Sclerotinia fructicola*.

Because a compound may act as a competitive inhibitor on one organism is no reason to believe that it acts as such in another. Usually, competitive inhibition is most striking in those cases where the organism is heterotrophic for the particular metabolite being antagonized.

A very striking example of the complexity of enzyme inhibition as a toxic mechanism is described by Woolley and Shaw (1951). They found that 2-azaadenine is particularly lethal to *Lactobacillus* species. They tested the purines, hypoxanthine, adenine, and xanthine as possible metabolites being antagonized, with the following results. 2-Azaadenine competitively inhibits hypoxanthine (Fig. 7.3). Its toxicity is largely

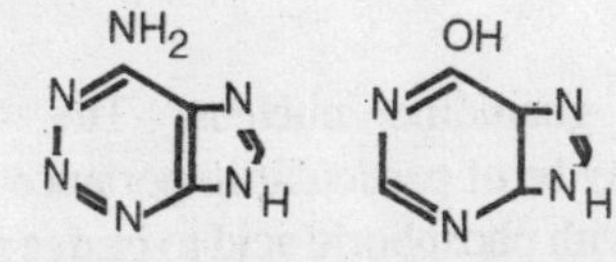


**Fig. 7.3.** Hypoxanthine (right) and its antimetabolite, 2-Azaadenine (left).

overcome by trace amounts of adenine, but is not appreciably alleviated by large amounts of xanthine. Without the inhibitors, xanthine is as effective as hypoxanthine in the purine nutrition of these bacteria. From their results, it becomes obvious that 2-azaadenine is highly specific in its effects on *Lactobacilli*. It is not simply a competitive inhibitor of purines as a class, but of one purine in particular, hypoxanthine.

Their experiments again serves to illustrate the physiological implications that may result from toxicological studies. Their results indicate that the chain of nucleoside synthesis in *Lactobacillus* goes from xanthine → hypoxanthine → adenine. If 2-azaadenine breaks the chain at hypoxanthine, the addition of more xanthine could have little effect. The addition of adenine, however, would be beyond the block and nucleic acid synthesis could continue.

Noncompetitive inhibition as a toxic mechanism may be postulated for almost any toxicant which can denature proteins. For example, picric acid, a protein precipitant, is quite fungitoxic and may act by denaturing proteins. The same may be said for formaldehyde. Any of the amino acid reagents shown to be fungitoxic by Horsfall and Zentmyer (1944) may be acting as noncompetitive inhibitors.

Uncompetitive inhibition has not been postulated as a toxic mechanism for any one particular known toxicant. This type of inhibition must surely contribute to the toxicity of some poisons.

Many known fungitoxicants probably act as nonspecific enzyme inhibitors. The enzyme proteins may lose their activity if the attacking toxic molecule has a particular avidity for, or ability to react with any of the reactive groups on the protein.

The susceptibility of enzymes may be realized when we consider the number of reactive sites possessed by amino acids which make up the proteins. Some of these reactive sites are primary amino groups (—$NH_2$), alcohol groups intermediate between aliphatic and aromatic (the —OH of hydroxyproline), the imino group (=NH), the acid amide group (—$CONH_2$), the sulfhydryl group (—SH), the α-hydrogen of tryptophan, and the guanidine nucleus. The —OH groups of hydroxyamino acids may be of particular importance as susceptible sites because they combine with phosphoric acid to bridge polypeptide chains.

There are other toxic mechanisms closely related to enzyme inhibition which may well be discussed here. One of these, a variation on the theme of competitive inhibition, is a phenomenon classified by Markham (1958) as a type of "lethal synthesis." In this particular mechanism, the imposter molecules are so like the normal substrate molecules that the toxicants move along the chain of synthesis and are actually incorporated into the final building blocks of living substance. The building blocks containing these imposter molecules are defective and cannot be utilized in a normal fashion by the organism. The organism shows toxic effects as a consequence.

Heinrich et. al. (1952) found that the protozoan *Tetrahymena geleii*, which is readily poisoned by 8-azaguanine, incorporates this aminopurine into its ribonucleic acid when grown on sublethal doses of the toxicant. The toxic effect of 8-azaguanine is readily overcome by the simultaneous addition of the nucleoside, guanosine.

Another example of this type is reported by Dunn and Smith (1957). They found that *Escherichia coli* made sensitive to poisoning by 5-bromouracil incorporates this halogenated pyrimidine into its deoxyribonucleic acid. The toxic effect of this compound cannot be overcome by the addition of thymine for which it is known to be an antimetabolite. *E. coli* is not inhibited by 5-bromouracil nor does it incorporate it unless the bacteria are made to require thymine by treatment with sulfanilamide.

To call this type of toxicity a form of competitive inhibition may not be strictly proper, as it may or may not be reversible. It is quite apparent, however, that in these cases there must be competition between normal substrate and imposter. Once the imposter molecules become incorporated into the nucleic acids the toxic molecules may or may not be replaceable.

2. *Enzyme Stimulation.* Not only may toxicants act by inhibiting enzymes, but they may also act by overstimulating enzymes. Such overstimulation may throw the normal cellular physiology completely out of balance. This abnormal unbalance may be analogous to an overdose of insulin in humans. Byrde et. al. (1956), with *in vitro* studies of enzymes from *Monilinia laxa*, showed that copper greatly stimulates the activity of DPN oxidase, while *o*-phenylphenol greatly stimulates cytochrome c

oxidase. Mahler et. al. (1955) suggest a plan to explain the burst of molecular oxygen released from systems containing metalloflavoprotein enzymes (EF-Fe) when they are acted upon by quinones. Normally the electron transfer goes DPNH → EF-Fe → Cytochrome c. With quinone in the system, the electrons may be diverted completely from EF-Fe to molecular oxygen rather than cytochrome c, thus uncoupling all subsequent phosphorylations.

## C. Chelation

Before presenting chelation as a toxic mechanism it would be well to know something about chelates.

Chelates are cyclic structures formed by the union of metallic atoms with inorganic or organic molecules or ions. The term comes from *chela*, the Greek name for crab's claw. One molecule of the nonmetallic portion of the complex is called a ligand. Compounds which chelate must be able to combine at more than one point with metals. A simple 2 point, or bidentate, ligand is pictured in (Fig. 7.4). Each 8-quinolinol molecule is

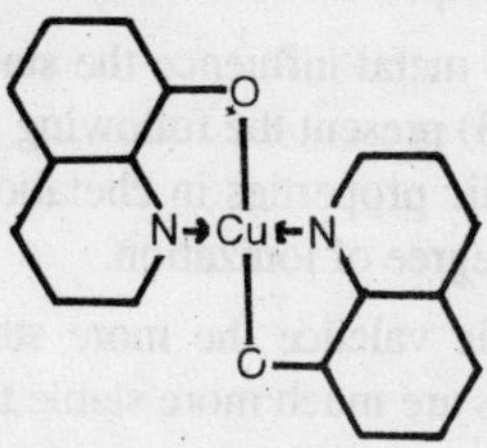


**Fig. 7.4.** 8-Quinolinol combining with copper. An example of bidentate ligands.

a ligand. Note that the metal is combined with each ligand by two bonds: (a) an ionic bonds, and (b) a coordinate bond. In a coordinate bond, both of the electrons shared by the combining atoms come from only one of the atoms, nitrogen in this case. Other chelates are known in which all the bonds between ligands and metal are coordinate, without any ionic bonding.

It is the multiple bonding feature of chelation which gives chelators their most interesting property. This property is the formation of an extremely stable complex which can hold metals so tightly that they are released or exchanged only with great difficulty.

The formation of chelate compounds is usually accompanied by color formation and the loss of polar properties by the ligand molecules. As the atoms on the ligand which combine with metals are usually nitrogen, oxygen, or sulfur, it is quite apparent why polar properties are lost. Whether or not the complex remains water-soluble depends on whether or not the ligand has polar groups not involved in chelation. For example, copper-8-quinolinolate is quite insoluble in water, while copper 8-quinolinol-5-sulfonate is water-soluble.

The smaller chelated cyclic structures are considered to be coplanar. According to the Baeyer strain theory, these, like other organic cyclic compounds, would be most stable when there are 5 or 6 atoms in each ring. Chelate cycles with less than 4 atoms or more than 7 atoms are known, but they are uncommon and usually unstable. Because of certain peculiarities of chelate rings, 4-membered structures are more common and more stable than 4-membered carbon rings. The probability of a compound being a chelator is highest if it has 2 electron-sharing atoms separated by 2 or 3 other atoms.

In the union of ligand with metal, the ligand is considered to be the electron-pair donor, or base, while the metal is the acceptor, or acid. All other things being equal, the greater the basic strength of the ligand, the more stable the chelate complexes it can form.

What properties of the metal influence the stability of the complex? Irving and Williams (1953) present the following three features as being the most important metallic properties in chelation: (a) valence, (b) the ionic radius, and (c) the degree of ionization.

The higher the cationic valence the more stable the complex. For example, $Fe^{+++}$ complexes are much more stable than $Fe^{++}$ complexes.

The larger the ionic radius the less stable the complex. $Mn^{++}$ has a much larger ionic radius than $Cu^{++}$ and usually forms less stable chelates.

Other things being equal, chelate stability is directly correlated with the second ionization potentials of the metals.

Chelating compounds are both common and important in biology. Among them are organic acids, amino acids, enzymes, vitamins, hemoglobin, and chlorophyll.

Chelation as a toxic mechanism has been a subject of active research and debate since it was first proposed by Zentmyer for fungi (1943) and by Albert for bacteria (1944).

Zentmyer (1944) showed that the toxicity of the powerful chelator 8-quinolinol (oxine) to the fungus can be overcome by the addition of excess zinc to the growth medium. Furthermore, reducing the pH of the medium to an acidity at which oxine cannot chelate, also destroyed its fungitoxicity. On this evidence, Zentmyer related the fungitoxicity of oxine to its ability to chelate, and proposed that oxine poisons by robbing the organism of required trace metals. Albert and his group were also able to destroy the antibacterial action of oxine by the addition of excess metal, iron in this case. In addition, they found that isomers or analogues of oxine able to chelate, were also antibacterial. Those related compounds unable to chelate, e.g., 2-quinolinol, were not antibacterial.

Soon, however, the metal-robbing theory was seriously questioned by Mason (1948) who showed that copper oxinate is more fungitoxic than the unchelated oxine. This was particularly damaging to the theory of metal-robbing. Since the copper chelate is one of the most stable of oxinates, it would not be expected to chelate other trace metals. Albert and his group found that oxine is not antibacterial unless iron is present in trace amounts. It must be made clear here that chelation as a toxic mechanism was not questioned. In doubt was only the theory that the toxicity of oxine derives from metals-robbing. It is ironic that the metal-robbing theory of toxicity should have been based on oxine which may or may not work in that way. There is not much doubt that chelators may exert their toxicity by metal-robbing. For example, Rich and Horsfall (1948) have shown that dimethylglyoxime will produce trace metal deficiency symptoms in *Aspergillus niger*. On dimethylglyoxime, the fungus grows in curds, rather than in a normal mat, and sporulation is poor or absent.

In what other way may chelators or chelate compounds poison? Albert and others have suggested that the poisonous form of oxine is the half chelate. Goksøyr (1955) also proposes the half chelate as the toxic species of dithiocarbamates. If this is so, then the half chelates may be poisoning by blocking functional groups. For example, an amino group and hydroxyl group properly spaced on the surface of an enzyme protein would form a mixed chelate with the half chelated toxicants. This is illustrated in (Fig. 7.5).

The same effect could be had with a sulfhydryl group in place of the amino or hydroxyl group.

Another type of functional group blocking would be the formation of mixed chelates with the phosphoric acid or nitrogen base portions of nucleotides. This may be illustrated (Fig. 7.6) with copper oxinate and adenosine-5′-diphosphate (ADP).

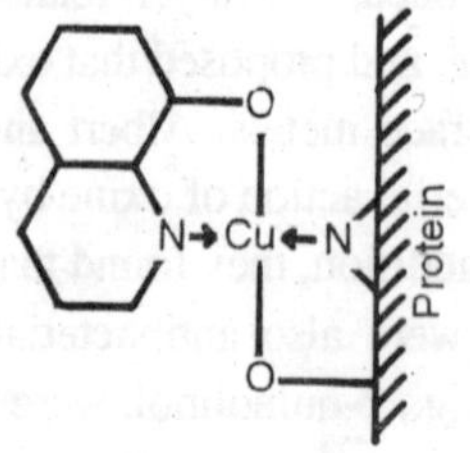


**Fig. 7.5.** Half chelate blocking functional groups.

So much for chelators and their possible toxic mechanisms. The fascinating stories of oxine and dithiocarbamates as toxicants will be described in detail in a later section.

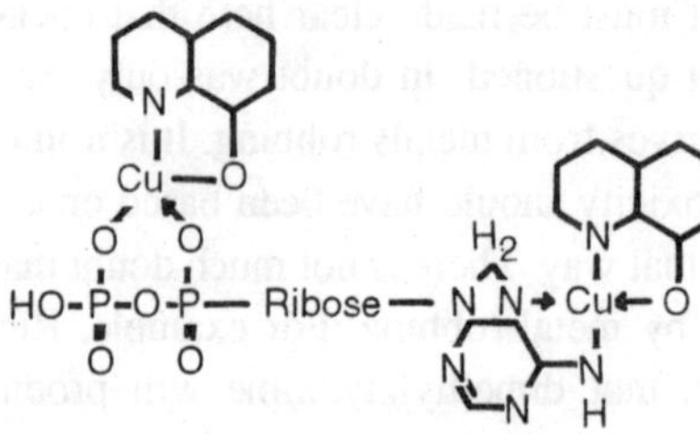


**Fig. 7.6.** Half chelate blocking ADP.

## D. Selective Solubility

Toxic compounds may act by selective solubility. Ferguson (1939) has shown that the insecticidal value of many chemically "inert" materials, such as saturated hydrocarbons, is correlated with their relative solubility in the medium in which they are applied. He pointed out that the partition of these compounds between the ambient medium and the biophase will be proportional to the degree of saturation of the ambient medium. In other words, the less soluble to toxicant in the applying medium, the greater the saturation of that medium and the greater the proportion entering the biophase. The degree of toxicity of these structurally non-specific compounds is directly proportional to their saturation of the medium in which they are applied. For example, nonspecific compounds *X*, *Y*, and *Z* may require respectively 1000 p.p.m., 100 p.p.m., and 1 p.p.m. to half-saturate the medium in which they are applied, whether it be water or air. These dosages should approximate half the maximum toxicity possible for treating organisms with compounds *X*, *Y*, and *Z*.

Ferguson was able to use this effect to explain another situation common in toxicity studies. Very often in comparing homologues of a toxic molecule, toxicity increases as the alkyl side chain lengthens. This continues until the addition of one more methyl group makes that homologue almost nontoxic. Ferguson suggests that the final methyl group would make the compound so insoluble in the applying fluid, that the amount required for complete saturation would partition less than a toxic dose into the biophase. This cut-off point will vary with compounds, applying medium, and tissue poisoned.

Ross and Ludwig (1957) have presented evidence that the fungi-toxicity of the *N-n*-alkylethylenethioureas follows the Ferguson effect.

All this is interesting and useful, but does not explain how supposedly "inert non-specific" compounds can poison. There are at least two possibilities.

The first is a solvent effect. If the inert molecules partially replace and force apart the biophase molecules, metabolic reaction rates could be altered enormously. Taking a test-tube example from organic chemistry, the reaction of ethyl iodide with triethylamine is only 1/1000 times as fast in hexane as it is in benzyl alcohol. The nonspecific toxicants could slow down vital reaction rates by reducing the frequency with which substrate molecules must collide with enzyme molecules essential for continuing normal processes.

The second way in which nonspecific toxicants may act is by altering the colloidal structure of the semipermeable membranes of the cell. Many of these toxicants have a particular avidity for the lipoproteins of the cellular membranes. They are also known to exert cellular narcosis, allowing cellular contents to leak from the cell. The most likely way that these compounds could do this is by affecting the structure of the hydrophilic colloids making up the semipermeable membranes.

### E. Dissolution

Toxicants may act by dissolving cellular components necessary to the physical structure of the cell. A good example is lysozyme, an antibiotic enzyme discovered by Fleming (1922) in mucous secretions. Lysozyme is a mucopolysaccharide depolymerase. It is capable of completely dissolving the cell walls of bacteria. Commercial lysozyme, prepared from egg whites, is assayed by its ability to dissolve suspensions of the bacterium *Micrococcus lysodeikticus*. The end-product of lysozyme activity is *N*-acetylglucosamine. There is no increase in reducing groups.

Lysozyme has been found in many bacteria, including those susceptible to its activity. It also occurs in *Ficus latex* and in commercial papain.

Rich has found that lysozyme, with no apparent effect on spore walls, will lyse the germ tubes of *Stemphylium sarcinaeforme*, allowing the cellular contents to escape. Rich also found that sodium hypochlorite causes the separation of the cells of this multicelled spore.

Carter and Lockwood (1957) described the lysis of the mycelium of *Glomerella cingulata* by a diffusate from *Streptomyces* sp. They also report the lysis of *G. cingulata* mycelium by antibiotics such as mycostatin and thiolutin, and by fungicides such as ferbam, thiram, and zineb. The latter effect they ascribe to a hastening of autolysis of the poisoned mycelium, rather than a direct dissolution of the mycelium by the fungicide.

A material may not cause a complete dissolution of cell wall, but may instead have a partial effect on its rigidity. Griseofulvin may act in this fashion. This antibiotic was found to be the "curling factor" produces by *Penicillium janczewskii*. It causes a spiral growth of the germ tubes of *Botrytis allii* and other fungi. This abnormality of the germ tubes apparently prevents the organism from developing mycelium.

**F. Precipitation**

Precipitation, the opposite of dissolution, may poison either by making metabolites unavailable or by destroying colloidal structure.

Polysulfides, prevent in lime sulfur, may produce their poisonous effect by precipitating essential metals and so making them unavailable. *Aspergillus niger*, grown in the presence of potassium sulfide, will produce yellow, rather than black, spores. This is a symptom of metal deficiency. Hydroxides or carbonates may also produce this effect.

Precipitation or coagulation of hydrophobic colloids, such as proteins, would also have a profound effect on normal cellular processes. The ability of a compound to bring about coagulation will depend on two things. First, the power to neutralize the charge on the dispersed phase, or the lowering of the zeta potential; and, second, the ability of the coagulant to take the stabilizing hydration water from the dispersed particles.

The ability to neutralize charges will depend on the valency of the ion charge oppositely to that of the dispersed phase. For example, if a protein sol is at a pH on the alkaline side of its isoelectric point, the dispersed phase will be negatively charged. The ability of the coagulant to reduce the charge on the protein micelle will be directly related to the valency of the cation of the coagulant. Actually, very small amounts of oppositely charged ions may cause a large drop in the zeta potential. The hydrophilic colloid may increase in viscosity as its charge is neutralized, but coagulation is prevented by the stabilizing effect of the hydrated shells around each micelle.

The final phase of coagulation depends on the ability of the toxicant to dehydrate. This ability to take water from the particles usually follows the lyotropic series. For the effect of anions to precipitate proteins, this series is usually shown as follows: citrate > tartrate > $SO_4$ > acetate > Cl > $NO_3$ > Br > I > CNS. Citrate shows the greatest effect and thiocyanate the least. This series for cations is, Th > Al > H > Sr > Ca > K > Na > Li. Ethanol and formaldehyde are two other toxicants which may act as protein coagulants.

### G. Hydrolysis

Acids or bases may poison by hydrolysis. Hydrolysis is the breaking of the reacting molecule and the combination of the pieces with $H^+$ and $OH^-$ from water.

From the viewpoint of toxicity, proteins are probably the most important biological compounds which may be hydrolyzed. Hydrolysis would, of course, be a denaturing process, particularly disruptive to enzymes. When enzymes are hydrolyzed, the most profound effect is the breaking of peptide linkages which join amino acids to make the peptide chains of which proteins are built (Equation 4).

$$H_2N{-}R{-}\overset{\overset{\displaystyle O}{\|}}{C}{-}\underset{\underset{\displaystyle H}{|}}{N}{-}R'{-}\overset{\overset{\displaystyle O}{\|}}{C}{-}OH + H_2O \xrightarrow[\text{base}]{\text{acid or}} H_2N{-}R{-}\overset{\overset{\displaystyle O}{\|}}{C}{-}OH + H_2N{-}R'{-}\overset{\overset{\displaystyle O}{\|}}{C}{-}OH \qquad \text{(Eq. 4)}$$

Mandels (1953) used weak acid treatment to destroy the activity of enzymes presumably located at or near the surface of *Myrothecium* spores.

Blood (1937) suggested that the acid production by fermenting tomato pulp is responsible for the elimination of *Corynebacterium michiganense* from tomato seed. He showed that soaking infested seed in acetic or lactic acid does as well as the fermentation process.

Tyner (1951) made the surprising discovery that merely soaking barely seed in water for a period of time can control loose smut. An investigation of this phenomenon by Leben et. al. (1956), and Jacquet et. al. (1957) showed that the control is achieved by organic acids. These organic acids, acetic, butyric, and formic, produced by bacteria operating during the water soaking, are the actual toxicants.

### H. Oxidation-Reduction

Oxidation-reduction changes, in order to cause any real damage to cells, must be drastic rather than mild. Living cells usually have too large a capacity to equilibrate changes in oxidation-reduction to expect any lasting toxicity from mild changes. Calcium hypochlorite, a strong oxidant, is very toxic. Whether or not it acts on a specific system is not important. It is probably nonspecific by virtue of its capacity to oxidize strongly many vital components within the cell. Ozone, another strong oxidant, is toxic to many types of cells. Its production by ultraviolet radiation may account in part for the deleterious effects of UV light. Such systems as $Cu^{++}$   $Cu^{++}$ may be poisonous by virtue of their oxidant action. For example, Albert (1952) reported that $Cu^{++}$ reacts rapidly with cysteine to give cystine and cuprous systeinate. Other toxic oxidizing agents are $SO_2$, $H_2O_2$, potassium permanganate, sodium perborate, and sodium dichromate. As will be discussed later, elemental sulfur may poison by acting as an oxidant.

Horsfall (1945) suggested that the dinitrophenols are toxic because they can oxidize and so split unsaturated fatty acids. Dinitrophenols are very effective against powdery mildews. Horsfall later (1956) proposed that these compounds kill conidia of powdery mildew by destroying the fatty envelope that retains the water these spores need to survive. The spores than die by desiccation.

Hydrogen sulfide is a powerful reducing agent particularly toxic to cells. It may act not only by precipitating metals, but also by holding them in a reduced state.

Oxidation-reduction systems may show definite antibacterial action not shown by either component of the system. Guest and Salle (1942) found that a combination of $FeCl_2$ with either $FeCl_3$ or $Fe_2(SO_4)_3$ is much more antibacterial than any of these compounds used separately. This was also true of Sn $Cl_2$ with Sn $Cl_4$, Fe $Cl_2$ with Sn $Cl_4$, Sn $Cl_2$ with Fe $Cl_3$, Mn $SO_4$ with Sn $Cl_4$, and Mn $SO_4$ with $Fe_2(SO_4)_3$. These heavy metal salts were most toxic when mixed in certain proportions. The increase in toxicity required that the mixture had one metal at a higher oxidation state than the other. Only the salts of heavy metals gave this effect.

## I. SURFACE ACTIVITY

Surface activity is a property of any substance whose presence in small amounts greatly alters the surface behavior of a given system. "Greatly alters" is rather a loose term, as most solutes will alter the surface behavior of the solvent to some extent. Although there are water-insoluble surfactants, we will be concerned with water-soluble organic compounds that markedly lower the surface tension of the liquid system in which they are dissolved. The most common of these surfactants is, of course, soap. A burgeoning chemical industry has produced a great umber of synthetic surfactants that have found uses in all phases of our daily life.

What are the molecular properties that make a surfactant? Basically, a surfactant molecule is a long molecule, one end polar and the other end nonpolar. The polar end has an affinity for polar substances, such, as water, while the nonpolar end has an affinity for nonpolar substances, such as grease, fats, and oils. When surfactant-containing water is applied to grease particles, the nonpolar end of the surfactant molecules is adsorbed by the grease. Since the polar end will not dissolve in the grease, the surfactants molecules stay at the surface. The surface-located polar ends having an affinity for water give the grease particle a miscibility with water. This allows the grease particles to become suspended in water, or emulsified, and so removed.

Surfactants lower the surface tension of water because the nonpolar ends of the surfactant molecules are forced out of water, and are concentrated and arranged linearly at right angles to the surface of the liquid, the polar end toward the liquid and the nonpolar end forced away from the liquid. The surface then acquires primarily the nonpolar properties of the surfactant. As the attraction between surfactant molecules is much less than the attraction of one water molecule for another water molecule, the surface tension is lowered. The reason why only a small amount of surfactant is required is that it is forced to the liquid-air or liquid-solid interface and all that is required for maximum effect is a unimolecular layer of the surfactant molecules at the interface.

There are three kinds of surfactants, anionic, cationic, and nonionic. The ionic surfactants ionize or dissociate into two parts, a large fatty radical and a small salt radical. If the charge in the fatty radical is negative, the compound is anionic. For example, a soap dissociates into $C_{17}H_{35}COO^-$ and $Na^{++}$, typical of an anionic surfactant. If the charge on the large radical is positive, the compound is classed as cationic. Cetylpyridinium chloride ionizes into

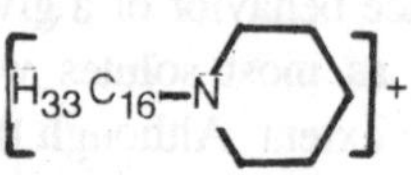

and Cl

making it a cationic surfactant. Nonionic, or ampholytic surfactants form zwitterions in solution. An example is cetylaminoacetic acid: $H_{33}C_{16}NH^+—COO^-$.

Soap is a week germicide. Its primary value as a sanitizer is to remove germs physically rather than to kill them. The real interest in the toxic and germicidal effects of surfactants dates from the work of Domagk (1935) who found that some cationic surfactants are powerful antibacterial compounds. Since then a number of quaternary ammonium compounds have been developed both as bactericides and fungicides. These have been principally alkylpyridinium halides and alkyl trimethylammonium halides. These cationic compounds are effective against both gram-positive and gram-negative bacteria and against fungi. Schwartz et. al. (1958) list 43 different types of cationic surfactant germicides.

Anionic surfactants are primarily active against gram-positive bacteria. Examples of anionic surfactant toxicants are soaps and certain aliphatic carboxylic acids.

Nonionic surfactants usually have little or no germicidal action. Schwartz et. al. (1958) list the salts of dioctylaminoethylglycine as examples of nonionic germicidal surfactants.

Surfactants appear to exert their toxic effects by breaking cross-links between polypeptide chains, or by disrupting colloidal structures. The effect on polypeptide structures is evident in the report of Sreenivasaya and Pirie (1938) who found that the protein of tobacco mosaic virus is disintegrated by surfactants. Anson (1939) also found that surfactants could split proteins into much smaller molecular units.

The profound effect of surfactants on colloidal structure is noted in toxicity studies by a disruption of the semipermeable membranes allowing the leaking materials from the cell. Horsfall (1956) discusses the importance of this as a toxic mechanism in both bacteria and fungi. Salton et. al. (1951) have taken electron micrographs of bacteria treated with cetyltrimethylammonium bromide and have observed morphological changes correlated with cell death. Bacteria treated with 45 mg. of cetyltrimethylammonium bromide per 1.5 mg. dry weight of bacteria for 5 minutes show a pulling away of the cytoplasmic membrane from the cell wall. This dosage of cetyltrimethylammonium bromide just initiates leaking. The proportion of cells (30%) killed by this treatment is the same as the number of treated cells showing contracted cytoplasm. Treating the cells with 900 mg. cetyltrimethylammonium bromide per 1.5 mg. of bacteria for 30 minutes strips off the cell walls.

### J. Osmotic Pressure Changes

Probably the oldest means of killing microorganisms is the preservation of foods with compounds that produce high osmotic pressures. Salting and sugaring are ancient means of preventing the growth of molds and bacteria on foods. Although the effective dosages of these materials are very high, salt and sugar were usually close at hand. On a dosage basis, this toxic mechanism is very inefficient, and will probably not be of importance in the development of future fungicides.

There are probably other toxic mechanisms of importance, but those listed here undoubtedly include those of principal interest.

It is entirely possible for a single toxicant to have more than one mode of action, one of which will be effective on one organism and another mode on a second organism. One can picture highly reactive molecules, such as captan, passing into the cell and reacting with metabolites *en route*. Toxicity or death of the cell results when any vital link is broken.

## II. INORGANIC FUNGITOXICANTS

### A. Metals

The most widely used inorganic toxicants are metallic compounds. Of the metals used as fungitoxicants, copper, specifically Bordeaux mixture, has been used in the largest amounts. Actually, in terms of minimum toxic dosages, inorganic metallic compounds are not nearly so toxic as a multitude of organic, nonmetallic toxicants. Even the organic metal complexes are more potent than the inorganic compounds containing the same metals.

According to Horsfall (1956), the descending order of fungitoxicity for the metal cations is approximately Ag > Hg > Cu > Cd → Cr > Ni > Pb > Co > Zn > Fe > Ca. McCallan and Wilcoxon (1934) found also that osmium and ruthenium are quite fungitoxic. Aluminum and palladium are only slightly toxic, and rhodium and iridium are essentially non-fungitoxic. McCallan and Wilcoxon (1934) studied the fungitoxicity of the elements in relation to their position on the periodic table. They generalized that toxicity within a group usually increased with increasing atomic weight. The principal exceptions are silver, much more toxic than gold, and chromium, usually more toxic than molybdenum and tungsten. A study of the periodic table shows that most toxic metals fall in groups I, II, and VIII.

Horsfall (1956) pointed out the good correlation between the toxicity of metals and their ability to form stable chelate compounds. His comparison of chelate stability with the toxicity of the metals is given below.

Toxicity : Hg > Cu > Cd > Ni > Pb > Co > Zn > Fe > Ca.

Chelate stability : Hg > Cu > Ni > Pb > Co = Zn > Cd > Fe > Mg.

Horsfall regards this as evidence that the primary toxic mechanisms of metals is their ability to combine with metabolites, enzymes, and proteins. Palladium and silver are two glaring exceptions to Horsfall's generalization. Palladium, only weakly toxic, forms very stable chelates, while silver, highly toxic, forms only weak chelates.

Shaw (1954) correlated toxicity with the insolubility of the metal sulfide. His diagrams show excellent correlations between the negative logarithm of lethal dose and the solubility product constants of the various metallic sulfides. This correlation was fairly consistent for the poisoning of diastase, urease, the paramecium, a planarium (*Polycelis nigra*), or a fish (the stickleback). Shaw believes this is the result of the "thermodynamic affinity of the inhibitor [metallic cations] for key functional group in the enzyme's catalytically active site [sulfhydryl groups]."

Shaw, of course, was preceded by Fildes (1940) who showed that mercury toxicity could be alleviated by sulfhydryl-containing compounds. Fildes suggested that mercury at low concentration acts by inactivating sulfhydryl groups without other permanent injury to the cell.

British anti-Lewisite (2, 4-dimercaptopropanol) was developed specifically as an antidote for the arsenical war-gas Lewisite (Cl CH=CH.As $Cl_2$). The development of BAL was based on the discovery that arsenic poisons by knocking out sulfhydryl groups.

Sulfhydryl compounds will also reverse copper toxicity of peas and of fungus spores. Foster showed that three kinds of seeds susceptible to copper injury contain sulfhydryl groups detectable with nitroprusside. Four kinds of seeds tolerant to copper contain no detectable sulfhydryl groups. Copper seed-treatments eliminated the detectable sulfhydryl groups of the copper-sensitive seeds. Both Yoder and Foster concluded that the sulfhydryl groups poisoned were essential to the functioning of the respiratory enzymes.

Metallic compounds are often used as astringents in medicine. Astringency is the precipitation of proteins by high concentrations of metallic compounds.

An obvious question is the relation of toxicity to the oxidation state of the metal. Apparently no generalization can be made concerning this point. Cupric ions are fungitoxic, but cuprous oxide is more toxic than cupric oxide. Mercuric ions are usually more toxic than mercurous ions.

No discussion of the toxicity of cations would be complete without including something about sorption. Are cations adsorbed at the periphery of the cell or are they absorbed freely into the interior? Bodnar and Terenyi (1930) found that the uptake of copper by *Tilletia tritici* followed the Langmuir adsorption isotherm. Sussman and Lowry (1955) have strong evidence that dormant ascospores of *Neurospora* also adsorb cations. McCallan (1948), working with copper sulfate on spores of *Alternaria oleraceae* and *Monilinia fructicola*, could not fit his data to Langmuir's isotherm. Miller et. al. (1953) do not believe that silver and cerium are merely adsorbed. As evidence, they cite the enormous quantities of cations taken up by their fungus spores. The number of cationic molecules taken up, if merely adsorbed, would form layers several molecules thick on the spore surface. In addition, a pretreatment with silver does not interfere with the uptake of cerium and vice versa. Interference would be expected if uptake of these cations were strictly by adsorption. The work of Owens and Miller (1957) also points to penetration of cations into spores with practically no concentration on the cell walls. Rothstein and Hayes (1956), however, find that yeast cells bind both univalent and bivalent cations at the cell periphery. The binding is rapid and reversible, with the bivalent ions being held more firmly than the univalent ions. They tentatively identify the binding sites as phosphoryl and carboxyl groups. These sites are apparently separated from the interior of the cell by a permeability barrier. Rothstein and Hayes suggest that certain cations will have special affinities for specific binding groups on the cell surface: $UO_2^{++}$ for carboxyl groups, $Hg^{++}$ for sulfhydryl groups, and $Zn^{++}$ for imidazole groups.

The evidence, then, is contradictory, but can be resolved. In organisms where there is a permeability barrier, adsorption is readily demonstrated. Where no permeability barrier exists or is destroyed by the toxicant, adsorption may play a role but it obscured by the free passage of cations into and out of the cell.

Another factor which confounds the sorption picture is the effect of metabolites leaking from the cell. Miller and McCallan (1957) have shown that treating spores of *M. fructicola* with toxic doses of $Ag^+$ for as little as 1.2 minutes increases their outward loss of phosphorous compounds. The loss of phosphorous compounds from these silver treated cells is 20 times that of the untreated spores. This mass of organic material capable of complexing will undoubtedly have a profound influence on the final location of cations on or in the spore.

### B. Nonmetals

1. *Sulfur*. Of the nonmetals, elemental sulfur is the most important fungitoxicant. In spite of its use as a pesticide for centuries, its method of toxic action is still a question for debate. Pollacci (1875) suggested that $H_2S$ produced from sulfur is actually responsible for the fungitoxic action of sulfur. Young (1922) proposed pentathionic acid as the toxic product from sulfur. Pentathionic acid results from the interaction of $H_2S$ and $SO_2$. The importance of pentathionic acid, $SO_2$, and $SO_3$ was largely discounted by Roach and Glynne (1928) and Wilcoxon and McCallan (1930). These workers found that the toxicity of these compounds is no more than could be accounted for by their hydrogen ion concentrations.

McCallan and Wilcoxon (1931) found that sulfur-sensitive spores produced large amounts of $H_2S$ from sulfur. They reverted to Pollacer's idea that $H_2S$, which they found to be enzymatically produced, was the cause of sulfur toxicity. When they again tested the $H_2S$ theory, Miller et. al. (1953) found that colloidal sulfur is as much as 50 times as toxic as $H_2S$ to sulfur-sensitive fungi. $H_2S$, they now believe, cannot be the toxic agent responsible for the toxicity of sulfur. They concluded that sulfur poisons because it is an active hydrogen acceptor and so interferes in normal hydrogenation reactions of the cell. The intensity of $H_2S$ production is a measure of the intensity of the hydrogenation process within the cell.

The theory of sulfur poisoning by competing for hydrogen seems plausible, but is still open to question. For example, methylene blue and ascorbic acid, both excellent hydrogen acceptors, are only weakly fungitoxic. Miller et. al. (1953) use as supporting evidence the work of Sciarini and Nord (1943) who concluded that, in *Fusaria*, elemental sulfur competes for hydrogen freed by dehydrogenases. Although Sciarini and Nord present such a conclusion, it is not supported by their data.

All of these ideas are possible. What Miller et. al (1953) have actually shown is that $H_2S$ produced within the spore is more toxic than $H_2S$ in the ambient fluid. This is not sufficient evidence to discard $H_2S$ as the source of sulfur toxicity. $H_2S$ which ionizes primarily as $HS^-$ may well poison metal-activated compounds just as metals poison sulfhydryl-activated compounds.

Miller et. al. (1953) also discount $H_2S$ as a factor because they find the $H_2S$ given off pretty well accounts for all the sulfur added to the system. However, the amount of $H_2S$ bound *in situ* by the micro-quantities of metals in the spore would be so small that it would be practically undetectable.

McCallan and Miller (1956) made the significant discovery that fungi in reducing sulfur release one mole of extra $CO_2$ for each mole of $H_2S$ produced. This contradicts Horsfall's (1956) suggestion that sulfur replaces oxygen in Krebs cycle intermediates. If this were so, then there should be less $CO_2$ than normal, rather than more. The production of $CO_2$ also indicates that elemental sulfur is not just competing for hydrogen, but appears to be acting as a strong oxidizing agent in the process of being reduced.

2. *Halogens*. Halogenated inorganic compounds are another group of useful toxicants. The commonest inorganic halogen compounds used as germicides are calcium or sodium hypochlorite. The toxicity of these compounds is directly correlated to the oxidizing power of their solutions. For example, Rideal and Evans (1921) showed that alkalinity

which reduces the germicidal effect of hypochlorites also reduces their oxidizing power. The strength of the hypochlorites is expressed as "available chlorine", a measure of oxidizing power.

CaOCl reacts with water and $CO_2$ from the air as in Equation 5.

$$CaOCl + HCO_3^- \rightarrow CaHCO_3 + ClO^-$$
$$ClO^- \rightarrow Cl^- + O \uparrow \quad \text{(Eq. 5)}$$

Chlorine evolved from acidification of CaOCl, will also oxidize as in Equation 6.

$$Cl_2 + H_2O \rightarrow HCl + HOCl$$
$$HOCl \rightarrow HCl + O \uparrow \quad \text{(Eq. 6)}$$

This is the method by which chlorine gas purifies water. Halogens may also act by halogenating metabolites, thus producing antimetabolites *in situ*.

3. *Formaldehyde*. Formaldehyde (HCHO) is widely used as a soil fumigant. Its action is primarily as a protein denaturant. The combination of formaldehyde with amino acids, as in the classic Sörenson formal titration (Equation 7), is representative of the way formaldehyde may combine with and denature proteins.

$$^{+}NH_3CH(R)COO^- + HCOH \rightarrow CH_2O.NH_2CH(R)COO^- + H \quad \text{(Eq. 7)}$$
(Zwitterion)

4. *Ammonia gas*. Ammonia fumigation has recently been used successfully in citrus storages for reducing fruit rots. Ammonia, highly soluble in water with which it forms $NH_4OH$, probably poisons fungi by drastic changes in their internal pH.

## III. ORGANIC FUNGITOXICANTS

### A. Dithiocarbamates

The greatest number of useful organic fungitoxicants has come from the dithiocarbamates. These compounds all have in common the grouping

$$(\!-\!\overset{|}{N}\!-\!\overset{\overset{S}{\|}}{C}\!-\!S\!-\!)$$

One series, based on the patent issued to Tisdale and Williams (1934), is made from the alkaline interaction of $CS_2$ and simple amines. These compounds are the dialkyldithiocarbamates and thiuram sulfides. The second series, based on Hester's patent (1943), is made from the alkaline interaction of $CS_2$ and ethylenediamine. These are the ethylene bisdithiocarbamates.

1. *Dialkyldithiocarbamates and Thiuram Sulfides.* The most widely used of these materials are zinc dimethyldithiocarbamate (ziram), ferric dimethyldithiocarbamate (ferbam), and tetramethylthiuram disulfide (thiram). Recently, another member of the family, sodium *N*-methyldithiocarbamate, has been finding use as a soil fungicide. The structures of these compounds are given in Fig. 7.7.

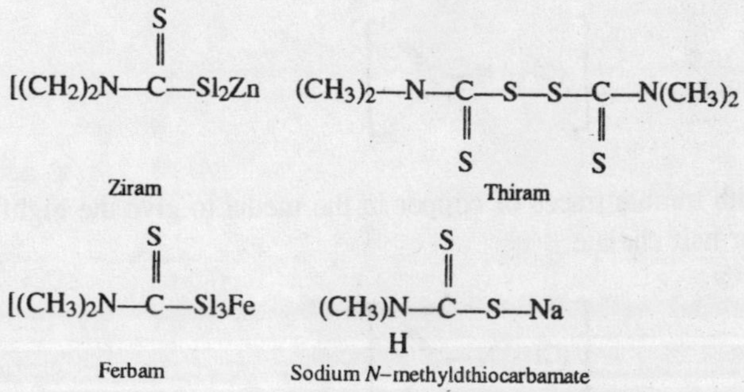


**Fig. 7.7.** Structures of ziram, thiram, ferbam, and sodium-*N*- methyldithiocarbamate.

The history of the development of these compounds and their performance in the laboratory has been reviewed thoroughly by Klöpping (1951), Horsfall (1956), and McCallan (1957).

The two most important clues to the toxic action of the dialkyl dithiocarbamates and the thiuram disulfides are (a) their formation of metal chelates and (b) their polymodal dosage response. The first property is a well-known one, commonly used in analytical chemistry. The second was first reported by Dimond et. al. (1941), for thiram. This latter effect, also called "inversion" and "quenching", is described in detail and logically explained by Goksøyr (1955).

Inversion of toxicity is pictured in Fig. 7.8.

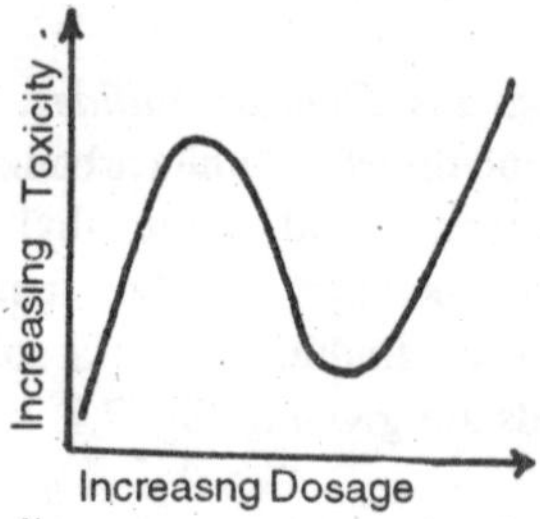


**Fig. 7.8.** Polymodal dosage response curve, or inversion of toxicity.

The zone of inversion, where increasing dosage actually depresses toxicity, has been particularly hard to explain. Goksøyr's explanation is as follows. The dimethyldithiocarbamate ion.

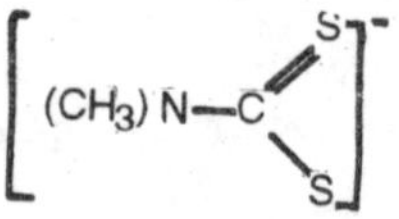

combines with minute traces of copper in the media to give the highly toxic 1 : 1, or half chelate,

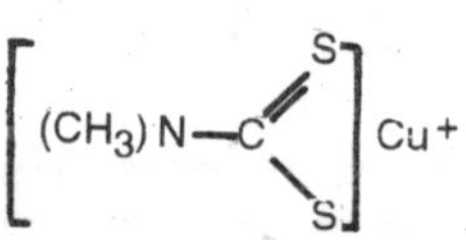

This is the toxic species. As the amount of copper in the media is constant, the addition of more dithiocarbamate forms the less toxic 1 : 2, or full chelate,

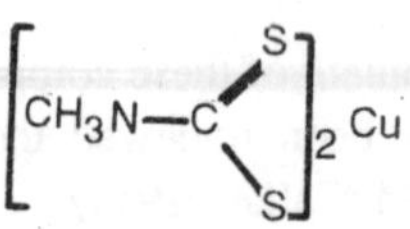

Toxcity falls. As more dithiocarbamate is added, the intrinsic toxicity of the dimethyldithiocarbamyl ion exerts itself, acting perhaps by binding essential metallic micronutrients. Thus toxicity increases again.

Goksøyr demonstrated that yeast cells reduce thiram to dimethyldithiocarbamyl ions. Hence, thiram activity and its polymodal dosage response curve is iden:ical to those of the dialkyldithiocarbamates.

As will be seen later, *N*-methyldithiocarbamate has an action more like the ethylenebisdithiocarbamates than like the dialkyldithiocarbamates.

2. *Bisdithiocarbamates.* This group of compounds arose around disodium ethylenebis (dithiocarbamate), now called nabam (Fig. 7.9).

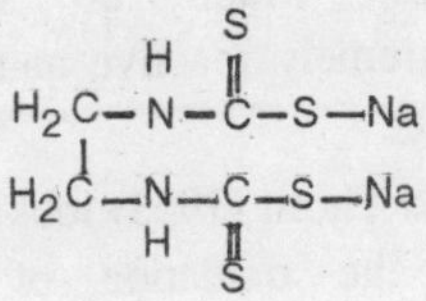

**Fig. 7.9.** Nabam.

The history of their development and the laboratory results of structure-activity studies are given in Barratt and Horsfall (1947), and Klöpping (1951).

What is the basis for their activity? Barratt and Horsfall (1947) proposed that since these compounds readily release $H_2S$, they poison by means of the $H_2S$ evolved. Rich and Horsfall (1950) demonstrated that $H_2S$ could not account for the toxicity of nabam. The clue to the currently accepted mechanism was noted in the title of the paper by Dimond et. al. (1943), in the first report on nabam. They called it a "water-soluble fungicide with tenacity." This anomaly puzzle them. In the data of Barratt and Horsfall appeared another clue. Zineb, the zinc salt of nabam, becomes more toxic upon aging. Klöpping (1951) blazed the trail which led to the present theory. Realizing that diisothiocyanates could form from bisdithiocarbamates, he tested the fungitoxicity of the diisothiocyanates. Klöpping's biological tests showed that the diisothiocyanates have the same biological spectrum as the bisdithiocarbamates. By "biological spectrum" he meant that different fungi have the same relative susceptibility to both types of compounds. In addition, the diisothiocyanates are much more toxic than the corresponding bisdithiocarbamates. Data from Klöpping (Table 7.1) are illustrative.

**TABLE 7.1**
**Fungitoxicity Data from Klöpping Illustrating Similarity between the "Biological Spectra" of Nabam and Ethylene Diisothiocyanate**

| Compound | *Minimum p.p.m. completely inhibiting growth* | | | |
|---|---|---|---|---|
| | *B. cinerea* | *P. italicum* | *A. niger* | *R. nigricans* |
| Nabam | 0.1 | 0.1 | 0.5 | 20.0 |
| Ethylenediisothicyanate | 0.05 | 0.02 | 0.05 | 10.0 |

Klöpping concluded that the ethylenebisdithiocarbamates act by releasing diisothiocyanates, which do the poisoning. The diisothiocyanates being extremely reactive and unstable, are hard to detect as breakdown products.

Shortly after, Ludwig and Thorn (1953) looking for insoluble (tenacious) compounds from the oxidation of nabam, found that ethylenethiuram monosulfide is a major product of nabam oxidation. They found that ethylenethiuram monosulfide (Fig. 7.10) is highly fungitoxic.

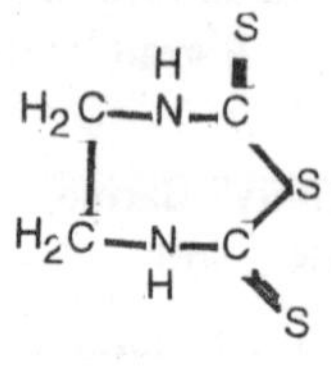

**Fig. 7.10.** Ethylenethiuram monosulfide.

Ethylenethiuram monosulfide they concluded to be the toxic breakdown product accounting for both the tenacity and toxicity of nabam. In a later paper, Ludwig et. al. (1955) reported that metallic ions catalyze the formation of isothiocyanates from ethylenethiuram monosulfide. They could demonstrate this only in a nonaqueous medium (chloroform).

In return, Sijpesteijn and van der Kerk (1954) confirmed the production of ethylenethiuram monosulfide from nabam.

The present theory of nabam action may be summarized from the literature. Freshly prepared, unoxidized nabam is presumed to be essentially nontoxic.

The acidic breakdown of nabam (Lopatecki and Newton, 1952) proceeds as in Equation 8.

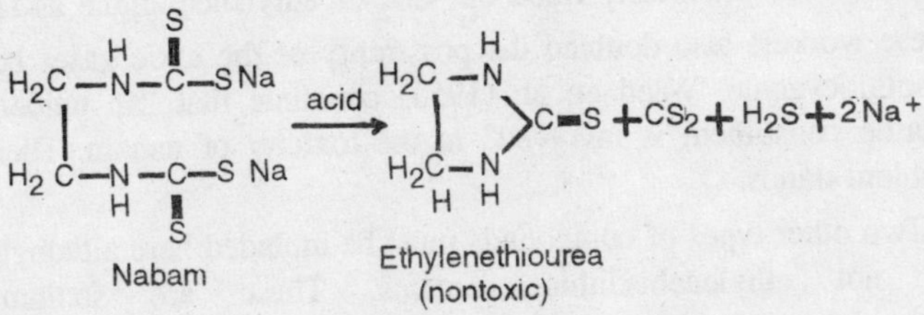


Basic hydrolysis and oxidation of nabam proceeds in another fashion (Equation 9).

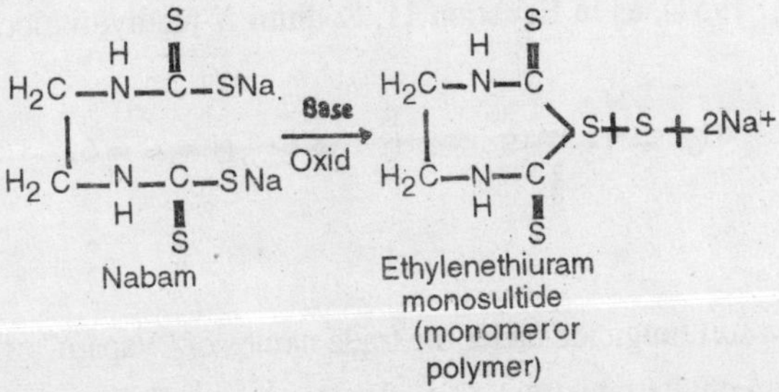


Ethylenethiuram monosulfide in turn breaks down to release ethylenediisothiocyanate (Equation 10).

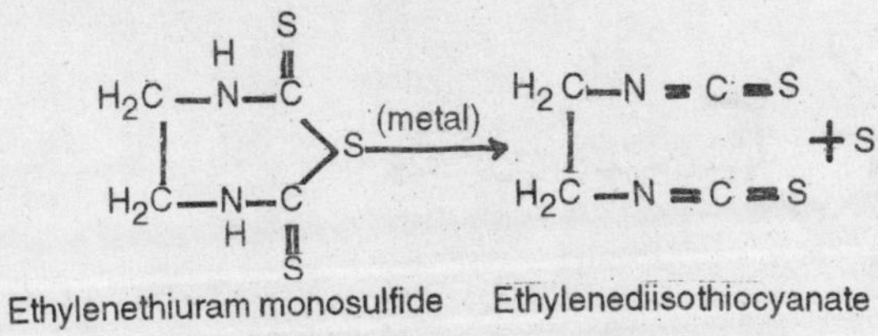


The diisothiocyanates, highly reactive and unstable, apparently combine with sulfhydryl groups to act as nonspecific poisons.

One of the unresolved problems of nabam toxicity is the nature of the gaseous toxicant which it evolves. This unidentified gas was first detected by Rich and Horsfall (1950), who demonstrated that the gas could not be $H_2S$ or $SO_2$. They did not consider $CS_2$ as a possibility because it is a relatively weak toxicant. In 1951, Cox et. al. showed that $CS_2$ and ethylenediamine evolve from nabam hydrolysis. They proposed that these two gases are the toxic vapors, recombining within the spore to give nabam. Lopatecki and Newton (1952) argued that the toxic vapors are simply $CS_2$ evolved from nabam. Later work on the problem by Weed et. al. (1953) practically ruled out $CS_2$ or ethylenediamine as factors. These workers also doubted the possibility of the toxic gases being a diisothiocyanate. Weed et. al. (1953) conclude that "an unidentified volatile constituent is involved" in the toxicity of nabam. There the problem stands.

Two other types of compounds must be included here although they are not ethylenebisdithiocarbamates. These are sodium *N*-methyldithiocarbamates and the rhodanines.

Sodium *N*-methyldithiocarbamate, in spite of structural similarity to the dialkyldithiocarbamates, is also a source of methylisothiocyanate (Klöpping, 1951), as in Equation 11. Sodium *N*-methyldithiocarbamate

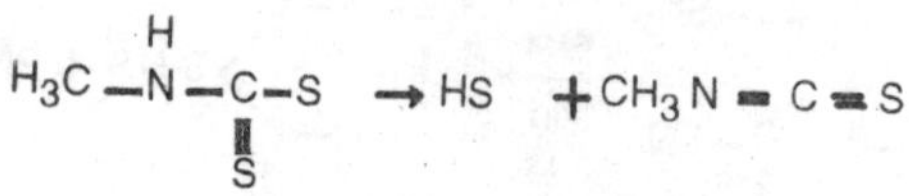

is used as a soil fungicide under the trade names of "Vapam" or "VPM."

3, 5-Dimethyltetrahydro-1, 3-5, 2H-thiadiazine-2-thione, the active ingredient of the soil fumigant "Mylone", also breaks down to give methylisothiocyanate as in Equation 12.

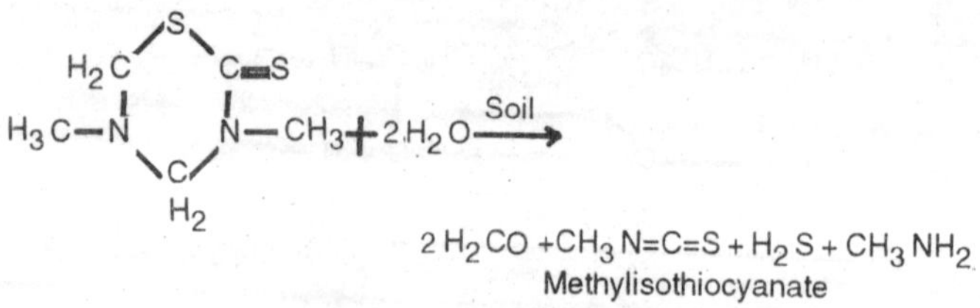

Another group of related fungitoxicants, the rhodanines, was described by van der Kerk et. al (1953). The structure-activity of the rhodanines was also reported in 1953 by Brown and her co-workers (1953). Van der Kerk reports (1956) that the rhodanines also are isothiocyanate fungicides and act according to Equation 13.

## B. Other Organic Sulfur Compounds

Although the dithiocarbamates have been the most useful and interesting of the organic sulfur fungicides, others also have been found to be fungitoxic.

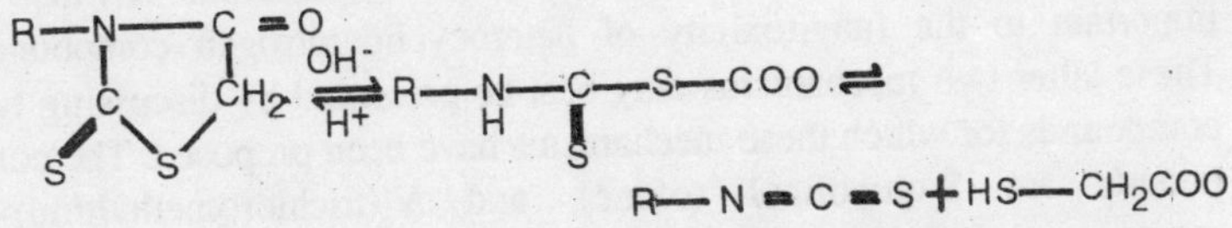

Some of these are the ethylenethioureas, the xanthates, 2-mercaptobenzothiazole, the thiobisphenols, and those toxicants having a trichloromethylthio group.

Rich and Horsfall (1954) noted that the *N*-alkylethlenethioureas increased in fungitoxicity with increasing length of alkyl chain. Fungitoxicity is maximal at the 8 carbon homologue and decreased as the chain lengthens beyond 8 carbons. Ross and Ludwig (1957) demonstrated that his series probably acts as physical toxicants exhibiting the Ferguson effect.

The xanthates, long used as disinfectants, were thoroughly investigate by Davies and Sexton (1946). They concluded that sulfur atoms on the C=S groups were not essential to the toxicity of xanthates. This conclusion was based on the essentially equal fungitoxicity of the following three compounds:

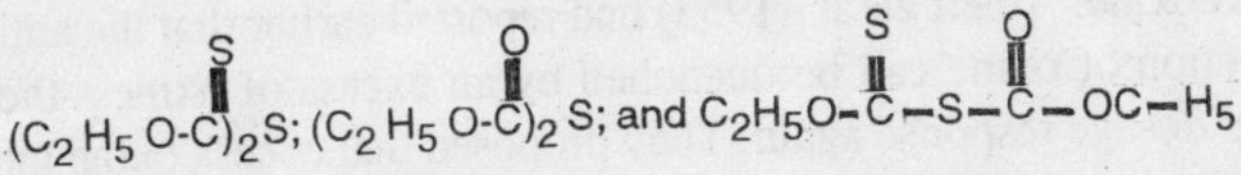

2-Mercaptobenzothiazole, although it is somewhat fungicidal, has found its principal use as a synergist with dialkyldithiocarbamates in proprietary mixtures sold as "Vancides."

The thiobisphenols will be discussed under phenolic fungicides. The trichloromethylthio compounds will be discussed under heterocyclic nitrogen compounds because of the interest in captan, the tetrahydrophthalimide derivative.

### C. Heterocyclic Nitrogen Compounds

Horsfall and Rich (1951) reported in the structure-activity relationships in a large group of heterocyclic nitrogen compounds. Many of these compounds are extremely fungitoxic. Their toxic mechanisms, however, differ widely. Certain of them, such as the ethylenethioureas just discussed, the nitrosopyrazoles, and the tetrahydropyrimidines, probably act as physical toxicants. Others, such as the 2-imidazolines have been reported to be competitive inhibitors of enzymes. The action, or acridine was discussed in an earlier section. Two other mechanisms may also be important to the fungitoxicity of heterocyclic nitrogen compounds. These latter two mechanisms may best be presented by discussing two compounds for which these mechanisms have been proposed. The compounds are 8-quinolinol (oxine), and *N*-(trichloromethylthio)-4-cyclohexene-1, 2-dicarboximide (captan). Oxine has been introduced in the previous section on chelation, and will be discussed first.

Oxine is the cornerstone for the theories of toxic mechanisms involving chelation. As mentioned earlier, the idea that oxine takes needed metals from organisms was called into doubt by evidence that the metal oxinates, particularly cupric oxinate, is more toxic than oxine itself. This evidence was further supported by Anderson and Swaby (1951), who found that oxine is a copper-deficient medium is only 1/48 as toxic as it is in a copper-sufficient medium. Copper must be present, therefore, for maximum fungitoxicity of oxine. Not only must there be copper present but the oxine must be able to form a chelate with it. For example, at a pH below which oxine chelates, it is a no longer toxic, regardless of the presence of copper. The formation of a copper oxinate, then, appears to be essential to the fungitoxicity of oxine.

Complications arose. Block (1957) showed that the fungitoxicity of copper oxinate can be reversed not only with excess copper but also with excess oxine. Albert et. al. (1953) had reported earlier that the antibacterial action of oxine can be quenched by an excess of oxine—the polymodal dosage response again. They proposed that excess oxine inhibited the formation within the cell of the 1 : 1, or half chelate which they suggested as the toxic form of oxine. Reversal by excess metal, they argued, was caused by inhibiting the 1 : 2 or full chelate in the ambient fluid. They pictured the process as follows. Oxine, because of water solubility, penetrates into cells best as the fully chelated metal complex which is less polar. Once inside the cell, the half chelated complex is required for toxicity. This half chelate may arise from the full chelate either by dissociation, or by the action of a hypothetical enzyme. Their theory is illustrated by Equation 14. Excess metal in the ambient fluid would drive the last equation to the left, reducing the amount of fully che-

$$\text{Oxine}^{-} + \text{Metal}^{++} \qquad \text{(Oxine) Metal}^{+}$$

Half Chelate
(Toxic within cell)

$$\text{(Oxine) Metal}^{+} + \text{Oxine} \qquad \text{(Oxine)}_2\ \text{Metal}$$

Full Chelate (Needed for permeation)

lated oxine needed for penetration. Excess oxine within the cell would drive the last equation to the right, reducing the concentration of the toxic species, the half chelate. Other chelators in excess will also cause reversal of oxine, ostensibly by competing for metals. Block (1955) demonstrated that in addition to the ability to chelate, analogues of oxine must also be lipoid-soluble to be fungitoxic.

There are at least two discrepancies in theory that should be noted. The first is that unchelated oxine enters cells quite readily. In fact, the theory for the reversal of toxicity by excess oxine depends on its entry into the cell. The second discrepancy is that washing with water will revive spores poisoned with oxine in a metal-deficient medium, but will not revive copper oxinate-treated spores. Washing with 0.1 *N* HCl is needed to revive the latter. These anomalies imply that unchelated oxine, which is toxic, acts in an area of the cell which does not require high lipoid-solubility to be entered. This action of oxine may well be metal-robbing. The chelated oxine acts at a site which requires high lipoid-solubility. This deeper site may well be within a microsome or the nucleus. The action of the half chelate may be the blocking of functional sites as suggested earlier.

Other workers have different views about the toxicity of oxine. Sexton (1953) considers it likely that oxine is fungitoxic because of its phenolic properties. Gale and Folkes (1957) find that oxine is one of the best inhibitors of the incorporation of glycine in disrupted staphylocolccal cells. They consider that the

portion of oxine structurally resembles, and competes with natural com-

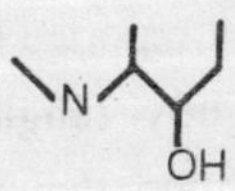

ponents of the system they were studying. They base this conclusion on two observations: (a) their system is stimulated by 6-amino-4-hydroxy derivatives of benzimidazole or benzotriazole; and (b) their system is not inhibited by the chelator ethylenediamine tetraacetic acid.

The next heterocyclic nitrogen compound to be discussed is captan (Fig. 7.11). The discussion of captan toxicity has centered on the importance of the —$SC(Cl)_3$ group. Included here, then, will be some analogues of captan that are not heterocyclic nitrogen compounds, but do have the —$SC(Cl)_3$ group.

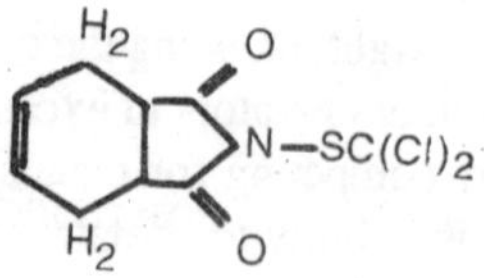

Fig. 7.11.

The fungitoxicity of captan was first reported by Kittleson (1952). He considered the fungitoxic group to be

$$>N—S—(C(Cl)_3$$

Horsfall and Rich (1953) presented evidence that —CO—NR—CO— could be the toxic grouping of captan. They based their conclusion on the similarity of the dosage-response slopes of captan and the highly fungitoxic antibiotic cycloheximide. Cycloheximide has the —CO—NH—CO— group but not —SC(Cl)3. Horsfall and Rich assigned to the trichlororomethylthio group the function of getting the tetrahydrophthalimide portion of captan into the cell. *N*-nitrosophthalimidine is another highly fungitoxic compound with the —CO—NR—CO— group but no —SC(Cl)3. The fungitoxicity of this compound was also used by Horsfall and Rich to support their theory. Horsfall and Rich (1954) pointed out that electronegative groups adjacent to a sulfur atom enhance activity. The electronegative carbonyl groups in captan would draw electrons from the nitrogen atom, making the N—S bond less stable.

Uhlenbroek and Koopmans (1957) found that $—SC(Cl)_3$ alone did not make a compound fungitoxic. They reported that $H_3CSSC(Cl)_3$ is non-fungitoxic, $C_6H_5SCC(Cl)_3$ is slightly fungitoxic, while $2\text{-}NO_2C_6H_4SSC(Cl)_3$ is very fungitoxic. In addition, they found that $—C(Cl)_3$ could be replaced without destroying fungitoxicity completely. For example, the ED50 of $H_3CC_6H_4SO_2SC(Cl)_3$ to *Fusarium culmorum* is in the range of 0.315 to 1.0 mg./liter, very fungitoxic. Replacing the $—C(Cl)_3$ with a $—CH_3$ results in a compound with an ED50 in the range 1.0–3.1 mg./liter, still quite fungitoxic. Unlenbroek and Koopmans found that the more electronegative groups on R in $RSSC(Cl)_3$, the more fungitoxic the analogue. In $RSO_2SC(Cl)_3$, however, the analogues are quite fungitoxic, regardless of the electronegativity of R. They concluded that the hypothesis of increased fungitoxicity by the electronegative groups adjacent to an S atom may apply for $RSSC(Cl)_3$ but not for $RSO_2SC(Cl)_3$. According to Ingold (1953), however, $—SO_2R$ is a very strong electronegative group in itself. Changes in R, therefore, may not give good evidence for or against the electronegativity theory, when $—SO_2R$ is adjacent to a sulfur atom.

Another theory of captan action was presented by Lukens and Sisler (1958). They found that captan interacts with sulfhydryl compounds, such as cysteine, to give thiophosgene

$$[Cl—\overset{\overset{\displaystyle S}{\|}}{C}—Cl]$$

as in Equation 15.

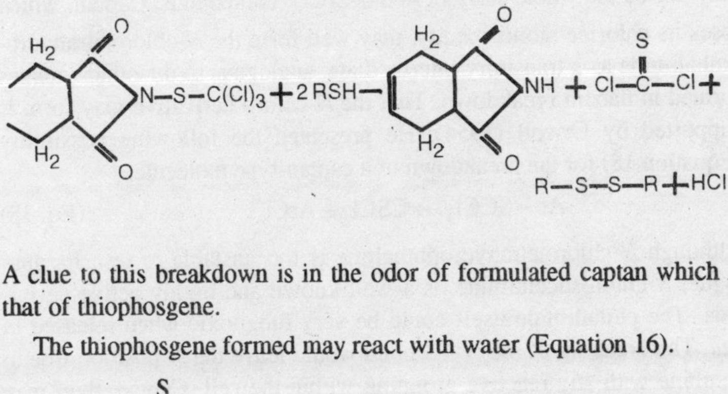

A clue to this breakdown is in the odor of formulated captan which is that of thiophosgene.

The thiophosgene formed may react with water (Equation 16).

$$\underset{\text{Thiophosgene}}{Cl—\overset{\overset{\displaystyle S}{\|}}{C}—Cl} + 2H_2O \rightarrow H_2S + CO_2 + 2HCl \qquad \text{(Eq. 16)}$$

Thiophosgene also polymerizes to give dithiophosgene which in turn may react with water (Equation 17).

$$\underset{\text{Dithiophosgene}}{Cl—\overset{\overset{\displaystyle S}{\|}}{C}—S—C(Cl)_3} + 2H_2O \rightarrow CS_2 + CO_2 + 4HCl \qquad \text{(Eq. 17)}$$

Lukens and Sisler proposed that the toxicity of captan is produced by the highly reactive thiophosgene resulting from the breakdown of captan by sulfhydryl groups. They were also able to demonstrate this breakdown by reacting captan with yeast cells. They concluded that essentially all of the toxicity of captan to yeast can be accounted for by the —S—C(Cl)$_3$ group. Thiophosgene produced *in situ* would ostensibly poison by combining with free sulfhydryl, amino, hydroxyl, and perhaps carboxyl groups. In addition, thiophosgene could form

bridges between properly spaced groups within the cell.

The production of thiophosgene is undoubtedly of importance to captan activity. Whether or not the captan acts simply as a thiophosgene producer remains doubtful. If, as Uhlenbroek and Koopmans (1957) reported, some —S—C(Cl)$_3$ compounds are nontoxic, and if replacing —C(Cl)$_3$ does not destroy fungitoxicity, then thiophosgene production may not be the whole story of —S—C(Cl)$_3$ compounds. Captan, which loses its chlorine atoms readily, may well form the *N*-chlorotetrahydrophthalimide as a transitory intermediate, analogous to diisothiocyanates formed in nabam breakdown. That the *N*-chloro derivative may form is supported by Orwoll (1954). He presented the following possibility (Equation 18) for the breakdown of a captan-type molecule.

$$Ar - SCCl_3 \rightarrow CSCl_2 + ArCl \qquad \text{(Eq. 18)}$$

Although *N*-chlorotetrahydrophthalmie is too unstable to test, its analogue, *N*-chlorosuccinimide, is a well known and highly active germicide. The phthalimide itself could be very fungitoxic when released *in situ*. The breaking of the N—S bond would leave the N position free to combine with any reactive grouping within the cell. Captan, then may poison by releasing highly reactive tetrahydrophthalimide radicals, thiophosgene, and chlorine.

So much for the heterocyclic nitrogen compounds, a particularly rich source of fungitoxicants.

### D. Quinones

Quinones are among the most powerful of the agricultural fungicides. The two most commonly used in practice are both chlorinated compounds. These are 1, 4-tetrachlorbenzoquinone (chloranil), and 2, 3-dichloro-1, 4-napthoquinone (dichlone).

Chloranil has found its use primarily as a seed protectant. As a foliage fungicide it usually fails to protect, being readily photolysed to nontoxic products.

Dichlone, much less readily photolysed, is widely used as a foliage spray. It has been particularly successful against apple scab. The potency of the quinones may be illustrated by the usage of dichlone for controlling apple scab. Most protective fungicides for apple scab control require 1 to 3 lb. of formulation per 100 gallons. Dichlone is used successfully at a dosage of 1/8 to 1/4 lb. per 100 gallons.

Owens (1953) points out a number of ways by which quinones may poison spores. As α, β-unsaturated compounds they may combine with sulfhydryl groups (Equation 19).

They may also interact with amino groups.

Quinones may also poison by rerouting electron transfer systems as described in an earlier section.

McNew and Burchfield (1951) found the following order of increasing fungitoxicity: anthraquinone < benzoquinone < phenanthraquinone < naphthoquinone. Halogenation, while decreasing water solubility and phytotoxicity, increases fungitoxicity. Chlorination improves fungitoxicity the most, bromination next, and iodination least. Alkylation ortho to

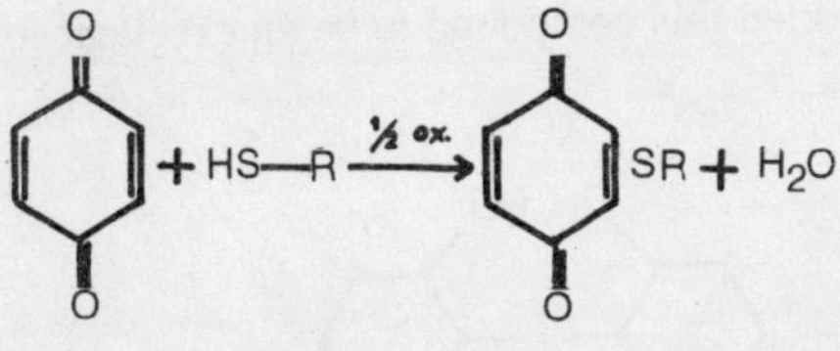


carbonyl groups reduces fungitoxicity, presumably by steric hindrance.

A point of interest is that the quinonoid structure may not be necessary to the activity of these compounds. For example, Owens (1953) reports than 2, 5-dichlorohydroquinone is almost as toxic as chloranil. Of course, data of Rich and Horsfall (1954) suggest that the hydroquinone may be oxidized to the quinone by fungi.

### E. Phenolic Compounds

Phenolic compounds have been of interest as germicides since the introduction of carbolic acid into surgery by Lister in the last century. Phenols as fungitoxicants or germicides have been particularly useful in medicine and industry rather than in agriculture. One of the most widely used germicides in medicinal preparations is hexylresorcinol. Various derivatives of cresol and phenylphenol are used as wood preservatives. Sodium dinitro-*o*-cresylate (Keitt, 1939) has been used as a spray on orchard floors to kill overwintering forms of fungi which cause fruit diseases.

One of the significant contributions to the structure-fungitoxicity relationships of phenols was made by Shirk (1954). He found that alkylating both positions ortho to the phenolic —OH greatly reduced fungitoxicity. In addition, *o*-chloro-6-alkylphenols are less fungitoxic than the corresponding *o*-alkylphenols. Shirk explains this effect as steric hindrance. He bases his suggestion on the report of Stillson et. al. (1945) who found that phenols substituted 2, 6 with large alkyl groups gave reactions not at all like typical phenols. These they called hindered phenols. Interfering with the reactivity of the phenolic —OH, then, profoundly affects fungitoxicity.

Another group of phenolic fungicides of interest are the bisphenols. The germicidal action of these compounds was first reported by Bechold and Ehrlich (1906). Marsh et. al. (1945), and Marsh and Butler (1946) found bisphenols to be highly effective mildew preventives on cloth. Their best material was 2, 2′-methylenebis (4-chlorophenol) also known as G-4, preventol GD and dichlorophene (Fig. 7.12) Goldsworthy and Gertler (1949) reported this compound to be an excellent foliage fungicide for peach trees.

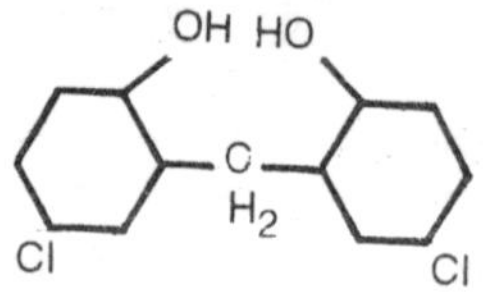


**Fig. 7.12.**

The corresponding thiobisphenols are also effective fungicides. Here, the methane bridge is replaced by a sulfur atom. Pfleger et. al. (1949), found that the thiobisphenols are highly toxic *in vitro* to fungi causing human diseases. Schraufstätter et. al. (1949) reported 2, 2′-thiobis (4-chlorophenol) as a particularly good antimycotic. Horsfall and Rich (1950) showed Schraufstätter's compound to be an excellent foliage protectant against apple scab. Unfortunately, it damaged the fruit.

Cade and Gump (1954) summarized the results of structure-activity research with bisphenols. Activity is increased by halogenation of each ring in the 4-positions. Any more halogenation decreases activity. The phenolic —OH's are most effective ortho to the bridge linking the rings.

The type of bridge between the two rings is not too important. The bridge must be small enough, however, to prevent spreading the hydroxyls too far apart. This latter evidence indicates that chelation may play a part in the toxicity of the bisphenols.

There are many other kinds of fungicides, such as salicylates and benzoates, that have been considered phenolic toxicants. The large group of phenolic toxicants are usually considered as nonspecific, protoplasmic poisons.

### F. Halogenated and Nitrated Aromatic Compounds

The halogenated aromatic compounds are usually toxic because they readily release nascent halogens. This is undoubtedly the mechanism of action of such compounds as choramine-T and *N*-chlorosuccinimide. The freed chlorine may act to form oxidizing $HClO^-$ as in the case of the hypochlorites. Another kind of action may be a direct halogenation of a metabolite to form an antimetabolite.

Nitrated aromatic compounds most potent as fungicides are usually either halogenated or they are phenols. One of these compounds, 2, 4-dinitro-*o*-cresol was mentioned in the previous section. Another is 2, 4-dinitrophenol, a powerful uncoupler of phosphorylation. Of the halogenated nitro-aromatics, one of the most powerful fungicides is 1-fluoro-3-bromo-4, 6-dinitrobenzene. The chlorinated nitrogenzenes have been of continuing interest since they were first described as fungitoxicants. Pentachloronitrobenzene and tetrachloronitrobenzene have found use both as soil fungicides and for the control of *Botrytis* sp. on plant parts. These compounds prevent sporulation, and are vapor-phase fungistats which do not prevent spores from germinating but do stop hyphal growth. Horsfall (1956) suggests that compounds of this type are antimitotics. It may be that pentachloronitrobenzene and tetrachloronitrobenzene act specifically as competitive inhibitors for inositol, a required growth substance for many fungi.

### G. Other Organic Fungitoxicants

The list of fungitoxic organic compounds is a long one. They cannot all be mentioned in one short chapter. Such fungitoxicants as quaternary ammonium compounds were mentioned in the section as surfactants. These have been tried for the control of plant diseases.

Not yet mentioned are the aliphatic fungitoxicants. These are of not too much importance and most probably act as nonspecific, physical toxicants. The organometallic fungicides have been of particular interest, so should not be omitted. Suffice it to say that many organometallic compounds, such as the organic mercurials, are more powerful than the inorganic compounds of the same metals. Of interest in this regard is tin, not very fungitoxic in inorganic combinations, but producing some highly fungitoxic organic compounds.

## IV. FASHIONING FUNGICIDES

### A. Success of Screening Programs

The artistry of the synthesizing chemists combined with the carefully designed screening programs of the biologists has given us a wealth of new and highly effective fungicides in recent years. The methods used for screening are as various as the laboratories so engaged. Although the selection of compounds to be screened is usually empirical, the tests to which the compounds are subjected are based on knowledge of plant diseases. Regardless of the method, there is very little question that most of the commercial fungicides now in use would have been selected by any of the presently operating screening programs. The screens work.

Concomitant with the selection of candidate fungicides, the screening programs have resulted in masses of structure-activity data. These data have been carefully scrutinized for "philosopher's stones" which will predict fungitoxic configurations. The literature abounds with such studies. These have been of use in predicting the activity of closely related compounds, and in a few cases have predicted the activity of untested configurations.

### B. Present Holes in Screening Programs

One of the main problems plaguing searchers for fungicides is the field failure of compounds which look promising in the laboratory. The present laboratory programs will pick all the winners, but, in the process, they also pick too many losers. Field testing is expensive.

Why do compounds look good in the laboratory and yet often are of no value for protecting plants in the field? Some of these factors involve the physical form of the fungicide. One of these physical factors of primary importance is particle size. A great number of studies have demonstrated the relationship between particle size and both fungitoxicity and tenacity. In order for fungitoxic compounds to succeed in the field, they must not only be in a proper physical form but also they must be carefully formulated. Burchfield and Goenaga (1957) have recently shown that even such an old and widely used fungicide as Bordeaux can be made more effective by formulation with certain wetting agents. Copper oxide succeeded as a foliage fungicide only after each particle was coated with gum arabic. The gum arabic coating prevented the agglomeration and consequent poor performance of the copper oxide particles. This made the difference between success and failure.

Very often candidate fungicides fail because they have not been subjected in the laboratory to the extremes of wealthering encountered in the field. Although laboratory testing ordinarily includes a washing or rain test, the compounds are not usually subjected to other weathering factors. Some of these additional factors are wind whipping of leaves, leaf expansion, high intensities of actinic light, long periods of continuous high humidity with attendant carbonation and oxidation, and periodic wetting and drying which would tend to increase hydrolysis. These factors, any of which may be critically important to field performance, are seldom or never tested in the laboratory. Fungitoxic compounds passing the rain test should be subjected to other weathering factors before being field tested.

Fungitoxicants may be lost not only by physical action but also through biological activity. The compounds may be degraded either by leaf emanations or by microbiological activity. Antibiotics are notoriously subject to degradation by soil microorganisms. It is interesting to note that biological degradation may actually make a compound more toxic. Byrde and Woodcock (1953) report such a case. They found that 1 : 4-diacetoxy-2 : 3-dichloronaphthalene, weakly toxic, becomes highly toxic when mixed with an enzyme preparation of high acetyl-esterase activity. Byrde and Woodcock suggest that the esterase changes this compound into the corresponding hydroquinone. Horsfall (1956) suggested that a phenol oxidase converts the hydroquinone to the quinone.

The pitfalls of a candidate fungicide are many and varied. It should be possible however, to devise laboratory tests to eliminate many of the compounds which now use up time, space, and money, failing in the field.

### C. Future of Fungicides

What is the future of fungicides? The future of fungicides is excellent. We become discouraged and go through periods of "agonizing reappraisal" only because our ignorance is so profound. We cannot make sense of the rapidly accumulating data. This chapter and others like it were written with the hope of illuminating signposts. Only by extrapolation from existing knowledge can we expect to find a "field theory" of fungicide action, if such exists. Without question, the many pathways pointed out by structure-activity studies eventually will become usable roads.

Should we continue to search for better fungicides? We need only to look to the history of insecticides for an affirmative answer. The discovery of the insecticidal properties of DDT, the chlorinated hydrocarbons, and the phosphates, has given us insecticides many times more powerful than the old. So it will be with fungicides.